Lecture Notes in Mathematics

2000

Editors:
J.-M. Morel, Cachan
F. Takens, Groningen
B. Teissier, Paris

T0184813

Harry Yserentant

Regularity and Approximability of Electronic Wave Functions

 Springer

Harry Yserentant
Technische Universität Berlin
Institut für Mathematik
Straße des 17. Juni 136
10623 Berlin
Germany
yserentant@math.tu-berlin.de

ISBN: 978-3-642-12247-7   e-ISBN: 978-3-642-12248-4
DOI: 10.1007/978-3-642-12248-4
Springer Heidelberg Dordrecht London New York

Lecture Notes in Mathematics ISSN print edition: 0075-8434
                                ISSN electronic edition: 1617-9692

Library of Congress Control Number: 2010927755

Mathematics Subject Classification (2000): 35J10, 35B65, 41A25, 41A63, 68Q17

Cover design: SPi Publisher Services

Printed on acid-free paper

springer.com

Preface

The electronic Schrödinger equation describes the motion of N electrons under Coulomb interaction forces in a field of clamped nuclei. Solutions of this equation depend on $3N$ variables, three spatial dimensions for each electron. Approximating the solutions is thus inordinately challenging, and it is conventionally believed that a reduction to simplified models, such as those of the Hartree-Fock method or density functional theory, is the only tenable approach. This book seeks to convince the reader that this conventional wisdom need not be ironclad: the regularity of the solutions, which increases with the number of electrons, the decay behavior of their mixed derivatives, and the antisymmetry enforced by the Pauli principle contribute properties that allow these functions to be approximated with an order of complexity which comes arbitrarily close to that for a system of one or two electrons. The present notes arose from lectures that I gave in Berlin during the academic year 2008/09 to introduce beginning graduate students of mathematics into this subject. They are kept on an intermediate level that should be accessible to an audience of this kind as well as to physicists and theoretical chemists with a corresponding mathematical training. The text requires a good knowledge of analysis to the extent taught at German universities in the first two years of study, including Lebesgue integration and some basic facts on Banach and Hilbert spaces (completion, orthogonality, projection theorem, Lax-Milgram theorem, weak convergence), but no deeper knowledge of the theory of partial differential equations, of functional analysis, or quantum theory. I thank everybody with whom I had the opportunity to discuss the topic during the past years, my coworkers both from Tübingen and Berlin, above all Jerry Gagelman, who read this text very carefully, found many inconsistencies, and to whom I owe many hints to improve my English, and particularly my colleagues Hanns Ruder, who raised my awareness of the physical background, and Reinhold Schneider, who generously shared all his knowledge and insight into quantum-chemical approximation methods. The Deutsche Forschungsgemeinschaft supported my work through several projects, inside and outside the DFG-Research Center MATHEON. I dedicate this book to my sons Klaus and Max.

Berlin, September 2009 *Harry Yserentant*

Contents

Chapter 1
Introduction and Outline

The approximation of high-dimensional functions, whether they be given explicitly or implicitly as solutions of differential equations, represents one of the grand challenges of applied mathematics. High-dimensional problems arise in many fields of application such as data analysis and statistics, but first of all in the sciences. One of the most notorious and complicated problems of this type is the Schrödinger equation. The Schrödinger equation forms the basis of quantum mechanics and is of fundamental importance for our understanding of atoms and molecules. It links chemistry to physics and describes a system of electrons and nuclei that interact by Coulomb attraction and repulsion forces. As proposed by Born and Oppenheimer in the nascency of quantum mechanics, the slower motion of the nuclei is mostly separated from that of the electrons. This results in the electronic Schrödinger equation, the problem to find the eigenvalues and eigenfunctions of the Hamilton operator

$$ H = -\frac{1}{2} \sum_{i=1}^{N} \Delta_i - \sum_{i=1}^{N} \sum_{\nu=1}^{K} \frac{Z_\nu}{|x_i - a_\nu|} + \frac{1}{2} \sum_{\substack{i,j=1 \\ i \neq j}}^{N} \frac{1}{|x_i - x_j|} \qquad (1.1) $$

written down here in dimensionless form or atomic units. It acts on functions with arguments $x_1, \ldots, x_N \in \mathbb{R}^3$, which are associated with the positions of the considered electrons. The a_ν are the fixed positions of the nuclei and the positive values Z_ν the charges of the nuclei in multiples of the absolute electron charge.

The mathematical theory of the Schrödinger equation for a system of charged particles is today a central, highly developed part of mathematical physics. Starting point was Kato's work [48] in which he showed that Hamilton operators of the given form fit into the abstract framework that was laid by von Neumann [64] a short time after Schrödinger [73] set up his equation and Born and Oppenheimer [11] simplified it. An important breakthrough was the Hunziker-van Winter-Zhislin theorem [46, 90, 98], which states that the spectrum of an atom or molecule consists of isolated eigenvalues $\lambda_0 \leq \lambda < \Sigma$ of finite multiplicity between a minimum eigenvalue λ_0 and a ionization bound Σ and an essential spectrum $\lambda \geq \Sigma$. The mathematical theory of the Schrödinger equation traditionally centers on spectral theory. Of at least equal importance in the given context are the regularity properties of the eigenfunctions, whose study began with [49]. For newer developments in this

H. Yserentant, *Regularity and Approximability of Electronic Wave Functions*,
Lecture Notes in Mathematics 2000, DOI 10.1007/978-3-642-12248-4_1,
© Springer-Verlag Berlin Heidelberg 2010

direction, see [32] and [45]. Surveys on the mathematical theory of Schrödinger operators and the quantum N-body problem in particular are given in the articles [47,75] and in the monograph [38].

Because of its high-dimensionality, it seems to be completely hopeless to attack the electronic Schrödinger equation directly. Dirac, one of the fathers of quantum theory, commented on this in [25] with the often quoted words, "the underlying physical laws necessary for the mathematical theory of a large part of physics and the whole of chemistry are thus completely known, and the difficulty is only that the exact application of these laws leads to equations much too complicated to be soluble." This situation has not changed much during the last eighty years, and depending on what one understands by soluble, it will never change. Dirac continued, "it therefore becomes desirable that approximate practical methods of applying quantum mechanics should be developed, which can lead to an explanation of the main features of complex atomic systems without too much computation." Physicists and chemists followed Dirac's advice and invented, during the previous decades, a whole raft of such methods of steadily increasing sophistication. The most prominent are the Hartree-Fock method that arose a short time after the advent of quantum mechanics, and its many variants, extensions, and successors, and the density functional based methods, that have been introduced in the sixties of the last century and are based on the observation that the ground state energy is completely determined by the electron density. These methods present themselves a challenge for mathematics. Lieb and Simon [60] proved the existence of a Hartree-Fock ground state and Lions [62] the existence of infinitely many solutions of the Hartree-Fock equations. The existence of solutions of the more general multiconfiguration Hartree-Fock equations was proven by Friesecke [33] and Lewin [59]. The singularities of the solutions of the Hartree-Fock equations at the positions of the nuclei have recently been studied by Flad, Schneider, and Schulze [31]. Schneider [72] gave an insightful interpretation and analysis of one of the most accurate methods of this type, the coupled cluster method. See [42] and the Nobel lectures of Kohn [51] and of Pople [66] for an overview on the present state of the art in quantum chemistry, and [16,56–58] for more mathematically oriented expositions.

The current methods are highly successful and are routinely applied in practice, so that the goals Dirac formulated eighty years ago are today widely reached. Nevertheless the situation is not very satisfying from the point of view of a mathematician. This is because the success of many of these methods can only be explained by clever intuition. In the end, most of these methods resemble more simplified models than true, unbiased discretizations and, at least from the practical point of view, do not allow for a systematic improvement of the approximations. That is why mathematicians, encouraged by the progress in the approximation of high-dimensional functions, have recently tried to find points of attack to treat the electronic Schrödinger equation directly. Promising tools are tensor product techniques as they are developed on a broad basis in Wolfgang Hackbusch's group at the Max Planck Institute in Leipzig. Beylkin and Mohlenkamp [10] apply such techniques to a reformulation of the electronic Schrödinger equation as an integral equation.

The present text intends to contribute to these developments. Aim is to identify structural properties of the electronic wave functions, the solutions of the electronic

Schrödinger equation, that will ideally enable breaking the curse of dimensionality. We start from ideas to the approximation of high-dimensional functions that emerged from the Russian school of numerical analysis and approximation theory [7, 53, 54, 76] and have since then been reinvented several times [23, 24, 97]. They are known under the name hyperbolic-cross or sparse-grid approximation. Since the work of Zenger [97], approaches of this kind have become increasingly popular in the numerical solution of partial differential equations. For a comprehensive survey of such techniques, see [15] and, as it regards their application to truly high-dimensional problems, [35]. Among the first papers in which direct application of such ideas has been tried for the Schrödinger equation are [34, 39, 43]. More recent attempts are [27, 28, 36, 37], and very recently the doctoral theses [40, 96]. The order of convergence that such methods can reach is limited since the involved basis functions do not align with the singularities caused by the electron-electron interaction [29, 30]. Nevertheless such methods have a high potential as our considerations will show, and be it only for the study of the complexity of electronic wave functions.

The principle behind these constructions can best be understood by means of a model problem, the L_2-approximation of functions $u : \mathbb{R}^d \to \mathbb{R}$ that are odd and 2π-periodic in every coordinate direction on the cube $Q = [-\pi, \pi]^d$ by tensor products

$$\phi(k, x) = \prod_{i=1}^{d} \phi_{k_i}(x_i) \tag{1.2}$$

of the one-dimensional trigonometric polynomials $\phi_1, \phi_2, \phi_3, \ldots$ given by

$$\phi_{k_i}(\xi) = \frac{1}{\sqrt{\pi}} \sin(k_i \xi). \tag{1.3}$$

Functions with the given properties that are square integrable over Q can be expanded into a multivariate Fourier series and possess therefore the representation

$$u(x) = \sum_{k} \widehat{u}(k) \phi(k, x), \quad \widehat{u}(k) = \int_{Q} u(x) \phi(k, x) \, dx, \tag{1.4}$$

where the sum extends over the multi-indices $k = (k_1, \ldots, k_d) \in \mathbb{N}^d$ and its convergence has to be understood in the L_2-sense. The speed of convergence of this series depends on that with which the expansion coefficients $\widehat{u}(k)$ decay. Assume, for example, that u is continuously differentiable, which implies that

$$|u|_1^2 = \sum_{i=1}^{d} \int_{Q} \left| \frac{\partial u}{\partial x_i} \right|^2 dx = \sum_{k} \left(\sum_{i=1}^{d} k_i^2 \right) |\widehat{u}(k)|^2 \tag{1.5}$$

remains bounded. Consider now the finite part u_ε of the series (1.4) that extends over the multi-indices k inside the ball of radius $1/\varepsilon$ around the origin, that is, for which

$$\sum_{i=1}^{d} k_i^2 < \frac{1}{\varepsilon^2}. \tag{1.6}$$

Due to the orthonormality of the functions (1.2), u_ε is the best approximation of u by a linear combination of the selected basis functions and satisfies the error estimate

$$\|u - u_\varepsilon\|_0^2 \leq \varepsilon^2 \sum_k \left(\sum_{i=1}^d k_i^2 \right) |\widehat{u}(k)|^2 = \varepsilon^2 |u|_1^2 \qquad (1.7)$$

in the L_2-norm. Unfortunately, the number of these basis functions grows like

$$\sim \frac{1}{\varepsilon^d} \qquad (1.8)$$

for ε tending to zero, which is out of every reach for higher space dimensions d. The situation changes if one does not fix the smoothness of the functions to be approximated, but let it increase with the dimension. Assume, to avoid technicalities, that u possesses corresponding partial derivatives and that these are continuous and set

$$|u|_{1,\text{mix}}^2 = \int_Q \left| \frac{\partial^d u}{\partial x_1 \ldots \partial x_d} \right|^2 dx = \sum_k \left(\prod_{i=1}^d k_i \right)^2 |\widehat{u}(k)|^2. \qquad (1.9)$$

Let u_ε^* be the function represented by the finite part of the series (1.4) that extends over the multi-indices k now not inside a ball but inside the hyperboloid given by

$$\prod_{i=1}^d k_i < \frac{1}{\varepsilon}. \qquad (1.10)$$

The L_2-error can then, by the same reasons as above, be estimated as

$$\|u - u_\varepsilon^*\|_0^2 \leq \varepsilon^2 \sum_k \left(\prod_{i=1}^d k_i \right)^2 |\widehat{u}(k)|^2 = \varepsilon^2 |u|_{1,\text{mix}}^2 \qquad (1.11)$$

and tends again like $\mathscr{O}(\varepsilon)$ to zero. The difference is that the dimension of the space spanned by the functions (1.2) for which (1.10) holds, now increases only like

$$\sim |\log \varepsilon|^{d-1} \varepsilon^{-1}. \qquad (1.12)$$

This shows that a comparatively slow growth of the smoothness can help to reduce the complexity substantially, an observation that forms the basis of sparse grid techniques. Due to the presence of the logarithmic term, the applicability of such methods is, however, still limited to moderate space dimensions.

Because of the Pauli principle, physically admissible wave functions have typical symmetry properties that will later be discussed in detail. Such symmetry properties represent a possibility to escape from this dilemma without forcing up the smoothness requirements further, a fact that has first been noted by Hackbusch [39] and is basic for the present work. Assume that the functions u to be approximated are antisymmetric with respect to the exchange of their variables, that is, that

$$u(Px) = \text{sign}(P)u(x) \qquad (1.13)$$

holds for all permutation matrices P. It is not astonishing that symmetry properties such as the given one are immediately reflected in the expansion (1.4). Let

$$\widetilde{\phi}(k,x) = \frac{1}{\sqrt{d!}} \sum_P \text{sign}(P)\phi(k,Px) \qquad (1.14)$$

denote the renormalized, antisymmetric parts of the functions (1.2), where the sums extend over the $d!$ permutation matrices P of order d. By means of the corresponding permutations π of the indices $1,\dots,d$, they can be written as determinants

$$\frac{1}{\sqrt{d!}} \sum_\pi \text{sign}(\pi) \prod_{i=1}^{d} \phi_{k_i}(x_{\pi(i)}) \qquad (1.15)$$

and easily evaluated in this way. For the functions u in the given symmetry class, many terms in the expansion (1.4) can be combined. It finally collapses into

$$u(x) = \sum_{k_1 > \dots > k_d} \left(u, \widetilde{\phi}(k, \cdot)\right) \widetilde{\phi}(k,x), \qquad (1.16)$$

where the expansion coefficients are the L_2-inner products of u with the corresponding functions (1.14). The number of basis functions needed to reach a given accuracy is reduced by more than the factor $d!$, a significant gain. It can be shown (see Chap. 8 for details) that the number of ordered sequences $k_1 > k_2 > \dots > k_d$ of natural numbers that satisfy the condition (1.10) and with that also the number of basis function (1.14) needed to reach the accuracy $\mathscr{O}(\varepsilon)$ does not increase faster than

$$\sim \frac{1}{\varepsilon^{1+\vartheta}}, \qquad (1.17)$$

independent of d, where $\vartheta > 0$ is an arbitrarily chosen small number. In cases such as the given one the rate of convergence in terms of the number of basis functions needed to reach a given accuracy becomes independent of the space dimension.

The present work is motivated by these observations. It has the aim to transfer these techniques from our simple model problem to the electronic Schrödinger equation and to establish a mathematically sound basis for the development of numerical approximation methods. One may wonder that this can work considering all the singularities in the Schrödinger equation. The deeper reason for that is that the terms of which the interaction potentials are composed depend only on the coordinates of one or two electrons. This and the symmetry properties enforced by the Pauli principle suffice to show that the admissible solutions of the electronic Schrödinger equation fit into the indicated framework.

The Pauli principle is a basic physical principle that is associated with the indistinguishability of electrons and is independent of the Schrödinger equation. It is of fundamental importance for the structure of matter. Electrons have an internal property called spin that behaves in many respects like angular momentum. Although spin does not explicitly appear in the electronic Schrödinger equation, it influences the structure of atoms and molecules decisively. The spin σ_i of an electron can attain the two values $\pm 1/2$. Correspondingly, the true wave functions are of the form

$$\psi : (\mathbb{R}^3)^N \times \{-1/2, 1/2\}^N \to \mathbb{R} : (x, \sigma) \to \psi(x, \sigma), \tag{1.18}$$

that is, depend not only on the positions x_i, but also on the spins σ_i of the electrons. The Pauli principle states that only those wave functions ψ are admissible that change their sign under a simultaneous exchange of the positions x_i and x_j and the spins σ_i and σ_j of two electrons i and j, i.e., are antisymmetric in the sense that

$$\psi(Px, P\sigma) = \text{sign}(P)\psi(x, \sigma) \tag{1.19}$$

holds for arbitrary simultaneous permutations $x \to Px$ and $\sigma \to P\sigma$ of the electron positions and spins. The Pauli principle forces the admissible wave functions to vanish where $x_i = x_j$ and $\sigma_i = \sigma_j$ for $i \neq j$. Thus the probability that two electrons i and j with the same spin meet is zero, a purely quantum mechanical effect. The admissible solutions of the electronic Schrödinger equation are those that are components

$$u : (\mathbb{R}^3)^N \to \mathbb{R} : x \to \psi(x, \sigma) \tag{1.20}$$

of an antisymmetric wave function (1.18). They are classified by the spin vector σ, being antisymmetric with respect to every permutation of the electrons that keeps σ fixed. We will discuss these interrelations in Chap. 4 and will study the different components (1.20) separately. Let σ be a spin vector that remains fixed throughout, and let I_- and I_+ be the sets of the indices i of the electrons with spin $\sigma_i = -1/2$ and $\sigma_i = +1/2$. To both index sets we assign a norm that can best be expressed in terms of the Fourier transforms of the considered functions and is given by

$$\|u\|_{\pm}^2 = \int \left(1 + \sum_{i=1}^{N} \left|\frac{\omega_i}{\Omega}\right|^2\right) \prod_{i \in I_{\pm}} \left(1 + \left|\frac{\omega_i}{\Omega}\right|^2\right) |\hat{u}(\omega)|^2 \, d\omega. \tag{1.21}$$

These two norms are combined to a norm that is defined by

$$\|u\|^2 = \|u\|_-^2 + \|u\|_+^2. \tag{1.22}$$

The momentum vectors $\omega_i \in \mathbb{R}^3$ form together the vector $\omega \in (\mathbb{R}^3)^N$. Their euclidean length is $|\omega_i|$. The quantity Ω fixes a characteristic length scale that will be discussed below. The norm given by (1.22) is related to the norm (1.9) and measures mixed derivatives whose order increases with the number of the electrons. It is first only defined for the functions in the space $\mathscr{D}(\sigma)$ of the infinitely differentiable functions u with compact support that are antisymmetric in the described sense but can be extended to the space $X^1(\sigma)$, the completion of $\mathscr{D}(\sigma)$ under this norm. The space $X^1(\sigma)$ is a subspace of the Sobolev space H^1 consisting of functions that possess high-order mixed weak derivatives.

Our first result, which originates in the papers [92, 94] of the author and will be proven in Chap. 6, is that the eigenfunctions u of the Schrödinger operator (1.1) of corresponding (anti-)symmetry for eigenvalues below the ionization threshold, i.e., the infimum of the essential spectrum, are contained in $X^1(\sigma)$. This means that they possess mixed weak derivatives whose order increases with the number of electrons. The norm (1.22) of these eigenfunctions can be explicitly estimated in terms of the L_2-norm of the eigenfunctions. If $\Omega \geq C\sqrt{N}\max(N,Z)$ is chosen

$$\|\|u\|\| \leq 2\sqrt{e}\,\|u\|_0 \tag{1.23}$$

holds, where Z denotes the total charge of the nuclei and C is a generic constant depending neither on the number of the electrons nor on the number, the position, nor the charge of the nuclei. Conversely, there is a minimum $\Omega \leq C\sqrt{N}\max(N,Z)$ such that (1.23) holds for all these eigenfunctions independent of the associated eigenvalue. There are hints that this Ω behaves like the square root of the ground state energy. The estimate (1.23) depends on the partial antisymmetry of the eigenfunctions, particularly on the fact that the admissible wave functions vanish at many of the singular points of the electron-electron interaction potential, everywhere where electrons with the same spin meet. Only small portions of the frequency domain thus contribute substantially to the admissible eigenfunctions. This remark can be quantified with help of the notion of hyperbolic crosses, hyperboloid-like regions in the momentum space that consist of those ω for which

$$\prod_{i \in I_-}\left(1 + \left|\frac{\omega_i}{\Omega}\right|^2\right) + \prod_{i \in I_+}\left(1 + \left|\frac{\omega_i}{\Omega}\right|^2\right) \leq \frac{1}{\varepsilon^2}, \tag{1.24}$$

with $\varepsilon > 0$ given. If u_ε denotes that part of the wave function whose Fourier transform coincides with that of u on this domain and vanishes outside of it, the H^1-error

$$\|u - u_\varepsilon\|_1 \leq \varepsilon\,\|\|u - u_\varepsilon\|\| \leq \varepsilon\,\|\|u\|\| \tag{1.25}$$

tends to zero like $\mathscr{O}(\varepsilon)$ with increasing size of the crosses. This estimate is a first counterpart to the estimate (1.11) in the analysis of our model problem.

These observations, however, do not suffice to break the curse of dimensionality. As is known from [20] and is proven in Chap. 5, the eigenfunctions u for eigenvalues λ below the infimum $\Sigma(\sigma)$ of the essential spectrum decay exponentially, the decay rate depending on the eigenfunction. Let $R > 0$ satisfy the estimate

$$\frac{1}{2R^2} < \frac{\Sigma(\sigma) - \lambda}{N}, \tag{1.26}$$

that is, let it be big enough compared to the size of the gap between λ and $\Sigma(\sigma)$, and define the correspondingly exponentially weighted eigenfunction as

$$\widetilde{u}(x) = \exp\left(\sum_{i=1}^{N}\left|\frac{x_i}{R}\right|\right)u(x). \tag{1.27}$$

The weighted eigenfunction $\widetilde{u}$ is then not only square integrable, as follows from [20], it also belongs to the space $X^1(\sigma)$ and moreover satisfies the estimate

$$\|\|\widetilde{u}\|\| \leq 2\sqrt{e}\,\|\widetilde{u}\|_0. \tag{1.28}$$

This is shown in Chap. 6 along with the proof of (1.23). The parameter Ω scaling the frequencies is the same as before, common to all eigenfunctions for eigenvalues below the essential spectrum. In the limit of R tending to infinity (1.28) reduces to the estimate (1.23). With that the corresponding mixed derivatives of the given eigenfunctions decay exponentially in the L_2-sense. The estimate relates the decay of the eigenfunctions in the position and the frequency space to one another, i.e., their spatial extension and the length scales on which they vary. Estimates like (1.23) and (1.28) are characteristic for products of three-dimensional orbitals. Our results show that the solutions of the full Schrödinger equation behave in the same way and justify in this sense the picture of atoms and molecules that we have in our minds.

Estimates like (1.28) have striking consequences for the approximability of electronic wave functions and limit the complexity of the quantum-mechanical N-body problem. The idea is to expand the eigenfunctions of the electronic Schrödinger operator (1.1) into products of the eigenfunctions of three-dimensional operators

$$-\Delta + V, \quad \lim_{|x|\to\infty} V(x) = +\infty, \tag{1.29}$$

like the Hamilton operator of the harmonic oscillator with a locally square integrable potential $V \geq 0$, tending to infinity for its argument tending to infinity. The essential spectrum of such operators is empty so that they possess a complete L_2-orthonormal system of eigenfunctions $\phi_1, \phi_2, \phi_3, \ldots$ for eigenvalues $0 < \lambda_1 \leq \lambda_2 \leq \ldots$. Every L_2-function $u : \mathbb{R}^{3N} \to \mathbb{R}$ can therefore be represented as L_2-convergent series

$$u(x) = \sum_{k\in\mathbb{N}^N}\widehat{u}(k)\prod_{i=1}^{N}\phi_{k_i}(x_i), \quad \widehat{u}(k) = \left(u, \prod_{i=1}^{N}\phi_{k_i}\right). \tag{1.30}$$

The speed of convergence of this expansion is examined in Chaps. 7 and 8 for the given eigenfunctions u of the Schrödinger operator (1.1) under the condition

$$V(x_i) \leq V_i^*(x)^2, \quad V_i^*(x) = \frac{\Lambda_0}{R}\exp\left(\left|\frac{x_i}{R}\right|\right), \tag{1.31}$$

limiting the growth of the potential V, with R the length scale from (1.26) describing the decay of the considered eigenfunctions and Λ_0 a constant basically independent of R. The result can again be best described in terms of a kind of norm estimate

$$\sum_k \left(\sum_{i=1}^N \frac{\lambda_{k_i}}{\Omega^2} \right) \left(\prod_{i \in I_-} \frac{\lambda_{k_i}}{\Omega^2} + \prod_{i \in I_+} \frac{\lambda_{k_i}}{\Omega^2} \right) |\widehat{u}(k)|^2 \leq 4\,(u, Wu), \qquad (1.32)$$

where the weight function $W = W_- + W_+$ is composed of the two parts

$$W_{\pm} = \left(1 + \sum_{i=1}^N \left| \frac{V_i^*}{\Omega} \right|^2 \right) \prod_{i \in I_{\pm}} \left(1 + \left| \frac{V_i^*}{\Omega} \right|^2 \right) \qquad (1.33)$$

and Ω chosen as in the estimates (1.23) or (1.28). Interestingly, the right hand side of this estimate solely depends on the decay behavior of the considered eigenfunction.

The crucial point is the appearance of the two products of the eigenvalues λ_k in the estimate (1.32). These products grow similar to factorials. The reason is that the eigenvalues λ_k of corresponding operators (1.29) increase polynomially like

$$\lambda_k \gtrsim k^{\alpha/3} \qquad (1.34)$$

for potentials that grow sufficiently fast, at least as fast as polynomials. The three comes from the fact that we start from an expansion into products of three-dimensional eigenfunctions. The constant $\alpha < 2$ is related to the growth behavior of the potential V. It can come arbitrarily close to $\alpha = 2$ for correspondingly chosen potentials. Let $\varepsilon > 0$ be given and consider the finite dimensional space that is spanned by the correspondingly antisymmetrized tensor products of the three-dimensional eigenfunctions ϕ_{k_i} for which the associated eigenvalues λ_{k_i} satisfy the estimate

$$\prod_{i \in I_-} \frac{\lambda_{k_i}}{\Omega^2} + \prod_{i \in I_+} \frac{\lambda_{k_i}}{\Omega^2} < \frac{1}{\varepsilon^2}. \qquad (1.35)$$

Let u_ε be the L_2-orthogonal projection of one of the given solutions u of the Schrödinger equation onto this space. Moreover, let

$$\|u\|^2 = \sum_k \left(\sum_{i=1}^N \frac{\lambda_{k_i}}{\Omega^2} \right) |\widehat{u}(k)|^2. \qquad (1.36)$$

Since u_ε is the part of the expansion (1.30) of u associated with the selected product functions, respectively the eigenvalues λ_{k_i} for which (1.35) holds,

$$\|u - u_\varepsilon\| \leq \varepsilon \, |||u - u_\varepsilon||| \leq \varepsilon \, |||u|||. \qquad (1.37)$$

As the norm given by (1.36) dominates the H^1-norm up to a rather harmless constant, this means that u_ε approximates the solution with an H^1-error of order ε if one lets ε tend to zero. The parameter ε determines the size of the hyperbolic crosses (1.35). Therefore only a very small portion of the product eigenfunctions substantially contributes to the considered wave functions and a surprisingly high rate of convergence, related to the space dimension $3N$, can be achieved.

One can even go a step further. Assume that the potential V in the three-dimensional operator (1.29) is rotationally symmetric. The eigenfunctions, now labeled by integers $n, \ell \geq 0$ and $|m| \leq \ell$, are then of the form

$$\phi_{n\ell m}(x) = \frac{1}{r} f_{n\ell}(r) Y_\ell^m(x), \quad r = |x|, \tag{1.38}$$

where the radial parts $f_{n\ell}$ as well as the assigned eigenvalues $\lambda_{n\ell}$ do not depend on the index m and the Y_ℓ^m are the spherical harmonics, functions that are homogeneous of degree zero and thus depend only on the angular part x/r of x. The L_2-orthogonal expansion (1.30) of a square integrable function $u : (\mathbb{R}^3)^N \to \mathbb{R}$ becomes then

$$u(x) = \sum_{n,\ell,m} \widehat{u}(n, \ell, m) \prod_{i=1}^N \phi_{n_i \ell_i m_i}(x_i), \tag{1.39}$$

where n, ℓ, and m are multi-indices here. Define now the L_2-orthogonal projections

$$(Q(\ell, m)u)(x) = \sum_n \widehat{u}(n, \ell, m) \prod_{i=1}^N \phi_{n_i \ell_i m_i}(x_i) \tag{1.40}$$

in which the angular parts are kept fixed and the sum extends only over the corresponding radial parts. These projections are in fact independent of the chosen three-dimensional operator and can be defined without recourse to the given eigenfunction expansion. They map the Sobolev space H^1 into itself. For all functions in H^1

$$\|u\|_1^2 = \sum_{\ell, m} \|Q(\ell, m)u\|_1^2, \tag{1.41}$$

as is shown in Chap. 9. The point is that for the eigenfunctions u of the electronic Schrödinger operator (1.1) of corresponding antisymmetry the expression

$$\sum_\ell \sum_m \left\{ \prod_{i \in I_-} (1 + \ell_i(\ell_i + 1)) + \prod_{i \in I_+} (1 + \ell_i(\ell_i + 1)) \right\} \|Q(\ell, m)u\|_1^2 \tag{1.42}$$

remains finite. This is another important consequence from the regularity theory from Chap. 6. It states that only few of the projections contribute significantly to an admissible solution of the electronic Schrödinger equation and estimates the speed of convergence of the expansion (1.39) in terms of the angular momentum quantum numbers ℓ_i. To reach an H^1-error of order $\mathcal{O}(\varepsilon)$ hence it suffices to restrict oneself to the contributions of the tensor products of eigenfunctions $\phi_{n_i \ell_i m_i}$ for which

$$\prod_{i \in I_-} \left(1 + \ell_i(\ell_i + 1)\right) + \prod_{i \in I_+} \left(1 + \ell_i(\ell_i + 1)\right) < \frac{1}{\varepsilon^2}, \tag{1.43}$$

$$\prod_{i \in I_-} \frac{\lambda_{n_i \ell_i}}{\Omega^2} + \prod_{i \in I_+} \frac{\lambda_{n_i \ell_i}}{\Omega^2} < \frac{1}{\varepsilon^2}, \tag{1.44}$$

provided the potential V is adapted as described to the considered eigenfunction. The condition (1.43) represents an additional selection principle that can help substantially reduce the number of the antisymmetrized tensor products of eigenfunctions that are needed to reach a given accuracy. The expansion into tensor products of Gauss functions forms an example for the efficacy of such measures.

The final result is truly surprising. Our estimates demonstrate that the rate of convergence expressed in terms of the number of correspondingly antisymmetrized tensor products of the three-dimensional eigenfunctions involved astonishingly does not deteriorate with the space dimension $3N$ or the number N of electrons. It is almost the same as that for a one-electron problem for the case that all electrons have the same spin, and almost the same as that for a problem with two electrons otherwise. What that means for the numerical solution of the Schrödinger equation is not clear so far, but our considerations show at least that the complexity of the quantum-mechanical N-body problem is much lower than generally believed.

Keeping the intended audience in mind, the exposition starts with a short chapter on Fourier analysis and spaces of weakly differentiable functions. The third chapter gives a short introduction to quantum mechanics that is tailored to the later needs. An interesting point for physicists and chemists might be that we start from the weak form of the Schrödinger equation, an approach that is common in the theory of partial differential equations but less in the given context. Chapter 4 deals with the electronic Schrödinger equation itself, formulates it precisely, and embeds it into a functional analytic framework. As indicated we consider the spin components of the eigenfunctions separately and do not exploit the symmetry properties of the problem to the maximum extent. This approach is enforced by the distinct regularity properties of the components. Chapter 5 contains a short introduction to some notions from spectral theory, that are rewritten here in terms of the bilinear forms underlying the weak form of the eigenvalue problem, and discusses the Rayleigh-Ritz method for the approximate calculation of the eigenvalues and eigenfunctions. We characterize the infimum of the essential spectrum in the spirit of Agmon [3] and Persson [65] and prove a simple but for us basic result on the exponential decay of the eigenfunctions. The Chaps. 6 to 9 form the core of this work. They contain a lot of unpublished material going far beyond [92] and [94]. The results we have just sketched are derived and proven there in detail.

Chapter 2
Fourier Analysis

Fourier analysis deals with the representation of functions as superpositions of plane waves, of spatial or spatial-temporal nature. It plays in many respects a decisive role in this work. The Schrödinger equation of a free particle is a wave equation whose solutions are superpositions of such plane waves with a particular dispersion relation. The abstract framework of quantum mechanics is reflected in this picture and can be motivated and derived from it. Fourier analysis plays moreover an extraordinarily important role in the mathematical analysis of partial differential equations like the Schrödinger equation and is basic for our considerations. We begin therefore with an elementary introduction to Fourier analysis. We start as usual from the Fourier transformation of rapidly decreasing functions that is then extended to integrable and square integrable functions. The third section of this chapter is devoted to the concept of weak derivative and its relation to Fourier analysis. We introduce rather general L_2-based spaces of weakly differentiable functions that include the usual isotropic Sobolev spaces but also spaces of functions with L_2-bounded mixed derivatives. Much more information on Fourier analysis can be found in monographs like [70] or [77], and on function spaces in [2, 85, 99].

2.1 Rapidly Decreasing Functions

A rapidly decreasing function $u : \mathbb{R}^n \to \mathbb{C}$, or in later chapters also from $\mathbb{R}^n$ to $\mathbb{R}$, is an infinitely differentiable function whose polynomially weighted partial derivatives

$$x \to x^\alpha (D^\beta u)(x) \tag{2.1}$$

remain bounded for all multi-indices $\alpha = (\alpha_1, \ldots, \alpha_n)$ and $\beta = (\beta_1, \ldots, \beta_n)$ with nonnegative integer components. Here we have used the known multi-index notation for the powers x^α of order $|\alpha| = \alpha_1 + \ldots + \alpha_n$ of the vector $x = (x_1, \ldots, x_n)$ in $\mathbb{R}^n$ and the partial derivatives. An example of such a function is the Gauss function

$$x \to \exp\left(-\frac{1}{2}|x|^2\right), \tag{2.2}$$

H. Yserentant, *Regularity and Approximability of Electronic Wave Functions*,
Lecture Notes in Mathematics 2000, DOI 10.1007/978-3-642-12248-4_2,
© Springer-Verlag Berlin Heidelberg 2010

where $|x|$ denotes the norm of the vector $x \in \mathbb{R}^n$ induced by the inner product

$$x \cdot y = \sum_{i=1}^{n} x_i y_i. \tag{2.3}$$

The rapidly decreasing functions form a complex vector space, the Schwartz space $\mathscr{S}$. The subspace $\mathscr{D}$ of $\mathscr{S}$ consists of the functions in $\mathscr{S}$ that have a compact support, that is, vanish outside bounded sets. The space $\mathscr{D}$ and with that also $\mathscr{S}$ are dense subspaces of the spaces L_1 and L_2 of integrable respectively square integrable complex-valued functions on the $\mathbb{R}^n$. This follows from the fact that the characteristic functions of axiparallel quadrilaterals, whose finite linear combinations are more or less by definition dense in L_1 and L_2, can be approximated arbitrarily well by functions in $\mathscr{D}$. Our strategy will be to work as far as possible with functions in $\mathscr{S}$ or even $\mathscr{D}$ and to transfer the corresponding results then by continuity arguments to their completions with respect to the considered norms. This begins with the definition of the Fourier transformation, first only for rapidly decreasing functions.

Definition 2.1. The Fourier transform of a rapidly decreasing function u is given by

$$\widehat{u}(\omega) = \left(\frac{1}{\sqrt{2\pi}}\right)^n \int u(x) e^{-i\omega \cdot x} dx \tag{2.4}$$

The Fourier integral (2.4) exists since rapidly decreasing functions are integrable. As an example we calculate the Fourier transform of the Gauss function (2.2).

Lemma 2.1. *The Fourier transform of the Gauss function (2.2) is*

$$\omega \rightarrow \exp\left(-\frac{1}{2}|\omega|^2\right). \tag{2.5}$$

Proof. The function (2.2) splits into a product of one-dimensional functions of same type. The Fourier transform of such a product is by Fubini's theorem the product of the one-dimensional Fourier transforms of these factors. We can therefore restrict ourselves to the case of one space dimension, to the function $f(x) = e^{-x^2/2}$ on $\mathbb{R}$. This function is the uniquely determined solution of the scalar initial value problem

$$f'(x) = -xf(x), \quad f(0) = 1.$$

Its Fourier transform has therefore the derivative

$$\widehat{f}'(\omega) = i\frac{1}{\sqrt{2\pi}} \int_{-\infty}^{\infty} f'(x) e^{-i\omega x} dx.$$

Integration by parts leads again to the differential equation

$$\widehat{f}'(\omega) = -\omega \widehat{f}(\omega).$$

Since the Fourier transform attains moreover at $\omega = 0$ the value

$$\widehat{f}(0) = \frac{1}{\sqrt{2\pi}} \int_{-\infty}^{\infty} e^{-x^2/2} \, dx = 1$$

and solves with that the same initial value problem as f, it coincides with f. □

The main reason to start with rapidly decreasing functions is the following:

Theorem 2.1. *The Fourier transform of a function in $\mathscr{S}$ is again rapidly decreasing.*

Proof. Since one is allowed to differentiate under the integral sign,

$$(i\omega)^\beta (D^\alpha \widehat{u})(\omega) = \left(\frac{1}{\sqrt{2\pi}}\right)^n \int u_\alpha(x) \, (i\omega)^\beta \, e^{-i\omega \cdot x} \, dx,$$

where $u_\alpha(x) = (-ix)^\alpha u(x)$ is again a rapidly decreasing function. Since

$$(i\omega)^\beta e^{-i\omega \cdot x} = (-1)^{|\beta|} D_x^\beta \{ e^{-i\omega \cdot x} \}$$

and as u_α and all partial derivatives of this function vanish sufficiently fast at infinity, Fubini's theorem and multiple integration by parts yield finally the representation

$$(i\omega)^\beta (D^\alpha \widehat{u})(\omega) = \left(\frac{1}{\sqrt{2\pi}}\right)^n \int (D^\beta u_\alpha)(x) e^{-i\omega \cdot x} \, dx$$

of the expression to be estimated. Since $x \to (D^\beta u_\alpha)(x)$ is as rapidly decreasing function integrable, the left hand side remains as required bounded in $\omega \in \mathbb{R}^n$. □

The Fourier transformation does therefore not lead out of the space of the rapidly decreasing functions, which is not the case for the functions in $\mathscr{D}$.

A fundamental property of the Fourier transformation is that functions can be recovered from their Fourier transforms by a very similar kind of transformation:

Theorem 2.2. *For all rapidly decreasing functions $u : \mathbb{R}^n \to \mathbb{C}$,*

$$u(x) = \left(\frac{1}{\sqrt{2\pi}}\right)^n \int \widehat{u}(\omega) e^{i\omega \cdot x} \, d\omega. \tag{2.6}$$

Proof. One cannot simply insert (2.4) into (2.6) and apply Fubini's theorem as one is then led to a diverging integral, in this form mathematical nonsense. We approximate the function u therefore first by the convolution integrals

$$(K_\vartheta * u)(x) = \int K_\vartheta(x - y) u(y) \, dy,$$

a kind of local averages, with the rescaled and normalized Gauss functions

$$K_\vartheta(x) = \left(\frac{1}{\sqrt{\vartheta}}\right)^n K\left(\frac{x}{\sqrt{\vartheta}}\right), \quad K(x) = \left(\frac{1}{\sqrt{2\pi}}\right)^n \exp\left(-\frac{1}{2}|x|^2\right),$$

as smoothing kernels. These kernels can with help of Lemma 2.1 be written as

$$K_\vartheta(x) = \left(\frac{1}{\sqrt{2\pi}}\right)^{2n} \int \exp\left(-\frac{\vartheta}{2}|\omega|^2\right) e^{i\omega\cdot x} d\omega,$$

a formula that can in view of (2.5) be already interpreted as a special case of (2.6). From Fubini's theorem and the definition of the Fourier transform therefore

$$(K_\vartheta * u)(x) = \left(\frac{1}{\sqrt{2\pi}}\right)^n \int \exp\left(-\frac{\vartheta}{2}|\omega|^2\right) \hat{u}(\omega) e^{i\omega\cdot x} d\omega$$

follows. Since $\hat{u}$ is integrable the right hand side of this equation converges by the dominated convergence theorem for ϑ tending to 0 to the right hand side of (2.6). To get the left hand side of (2.6), one rewrites the convolution integrals as

$$(K_\vartheta * u)(x) = \int K(y) u(x + \sqrt{\vartheta} y) dy.$$

Since u is as rapidly decreasing function bounded and continuous and since the Gauss function K is integrable, the dominated convergence theorem leads to

$$\lim_{\vartheta \to 0+} (K_\vartheta * u)(x) = \int K(y) u(x) dy = u(x).$$

This completes the proof of the inversion formula (2.6). $\qquad\square$

The Fourier inversion formula (2.6) shows that every rapidly decreasing function can be represented as Fourier transform of another rapidly decreasing function, as

$$u(x) = \left(\frac{1}{\sqrt{2\pi}}\right)^n \int \hat{u}(-\omega) e^{-i\omega\cdot x} d\omega. \tag{2.7}$$

The Fourier transformation is therefore a one-to-one mapping from the space of the rapidly decreasing functions to itself. Every rapidly decreasing function can in this sense be represented as superposition of plane waves.

Another consequence of the Fourier inversion theorem is the Plancherel theorem, often also denoted as Parseval identity in analogy to the corresponding property of Fourier series. It belongs undoubtedly to the central results of Fourier analysis.

Theorem 2.3. *For all rapidly decreasing functions u and v,*

$$\int \hat{u}(\omega) \overline{\hat{v}(\omega)} \, d\omega = \int u(x) \overline{v(x)} \, dx. \tag{2.8}$$

Proof. By the definition (2.4) of the Fourier transform of u,

$$\int \widehat{u}(\omega) \overline{\widehat{v}(\omega)} \, d\omega \, = \, \left(\frac{1}{\sqrt{2\pi}}\right)^n \iint u(x) \overline{\widehat{v}(\omega)} \, e^{-i\omega \cdot x} \, dx \, d\omega.$$

The Fourier inversion formula (2.6) applied to v leads conversely to

$$\int u(x) \overline{v(x)} \, dx \, = \, \left(\frac{1}{\sqrt{2\pi}}\right)^n \iint u(x) \overline{\widehat{v}(\omega)} \, e^{-i\omega \cdot x} \, d\omega \, dx.$$

The proposition follows from Fubini's theorem. □

The Plancherel theorem shows particularly that the Fourier transformation preserves the L_2-norm of a rapidly decreasing function, that is, that for all such functions u

$$\int |\widehat{u}(\omega)|^2 \, d\omega \, = \, \int |u(x)|^2 \, dx. \tag{2.9}$$

In other words, the Fourier transformation is a unitary mapping. This property is the key to the definition of the Fourier transform of arbitrary square integrable functions.

Besides its obvious physical meaning, a main reason to introduce the Fourier transformation is that it transforms derivatives to simple multiplications by polynomials. This follows differentiating the Fourier inversion formula, that is, from

$$(D^\alpha u)(x) \, = \, \left(\frac{1}{\sqrt{2\pi}}\right)^n \int (i\omega)^\alpha \, \widehat{u}(\omega) \, e^{i\omega \cdot x} \, d\omega, \tag{2.10}$$

and the one-to-one relation between a function and its Fourier transform.

Theorem 2.4. *The Fourier transforms of a rapidly decreasing function u and of its partial derivatives of arbitrary order are connected via the relation*

$$(\widehat{D^\alpha u})(\omega) \, = \, (i\omega)^\alpha \, \widehat{u}(\omega). \tag{2.11}$$

The relation (2.11) allows it to transform differential equations with constant coefficients to algebraic equations, and (2.10) offers a possibility to generalize the notion of derivative. This idea will be taken up in the next but one section.

2.2 Integrable and Square Integrable Functions

A much larger space than the space $\mathscr{S}$ of the rapidly decreasing functions and the natural domain of definition of the Fourier transformation is the space L_1, the space of the integrable functions $u : \mathbb{R}^n \to \mathbb{C}$, the measurable functions with finite L_1-norm

$$\|u\|_{L_1} \, = \, \int |u(x)| \, dx. \tag{2.12}$$

The rapidly decreasing functions and even the functions in $\mathscr{D}$, the space of the infinitely differentiable functions with compact support, form dense subsets of L_1.

Definition 2.2. The Fourier transform of a function u in L_1 is given by

$$\widehat{u}(\omega) = \left(\frac{1}{\sqrt{2\pi}}\right)^n \int u(x) e^{-i\omega \cdot x} dx. \tag{2.13}$$

The Fourier transform of an integrable function does not need to be itself integrable, which causes considerable difficulties and is one of the main reasons to start instead with the rapidly decreasing functions. However:

Theorem 2.5. *The Fourier transform of an integrable function is uniformly continuous and tends uniformly to zero for its argument tending to infinity.*

Proof. Let $u \in L_1$ and $u_1, u_2, \ldots$ be a sequence of rapidly decreasing functions with

$$|\widehat{u}(\omega) - \widehat{u}_k(\omega)| \leq \|u_k - u\|_{L_1} \to 0$$

for all $\omega \in \mathbb{R}^n$, that is, whose Fourier transforms converge uniformly to the Fourier transform of u. Since the $\widehat{u}_k$ are as rapidly decreasing functions uniformly continuous, the limit function $\widehat{u}$ is uniformly continuous, too. To prove the second assertion, we fix an $\varepsilon > 0$ and choose a sufficiently large index k, such that

$$|\widehat{u}(\omega)| \leq |\widehat{u}_k(\omega)| + \|u_k - u\|_{L_1} \leq |\widehat{u}_k(\omega)| + \varepsilon/2.$$

As $\widehat{u}_k$ is rapidly decreasing, there is an $R > 0$ with $|\widehat{u}_k(\omega)| < \varepsilon/2$ for $|\omega| > R$. For these ω, $|\widehat{u}(\omega)| < \varepsilon$, so that the function values $\widehat{u}(\omega)$ tend uniformly to zero.   □

The fact that the Fourier transform $\widehat{u}(\omega)$ of an integrable function u tends to zero for ω tending to infinity is usually denoted as the Riemann-Lebesgue theorem.

The Hilbert space L_2 consists of the square integrable functions u from $\mathbb{R}^n$ to $\mathbb{C}$, the measurable functions for which the L_2-norm given by the integral expression

$$\|u\|_0^2 = \int |u(x)|^2 dx \tag{2.14}$$

remains finite. Square integrable functions do not need to be integrable. The Fourier transform of such a function can therefore not simply be defined by the integral expression above. The Plancherel theorem offers a remedy. It shows that the Fourier transformation $u \to \widehat{u}$ can be uniquely extended from the dense subspace $\mathscr{S}$ of L_2 to a norm preserving, unitary linear mapping $F : L_2 \to L_2$. We define this mapping as the Fourier transformation on the space L_2 of the square integrable functions.

Theorem 2.6. *The L_2-Fourier transformation $F : L_2 \to L_2$ is a bijective, unitary linear mapping. If S denotes the reflection operator $u(x) \to u(-x)$, its inverse is*

$$F^{-1} = SF. \tag{2.15}$$

Proof. By the Fourier inversion formula (2.6) and (2.7), $u = SFFu = FSFu$ for all rapidly decreasing functions u. As F and S are bounded linear operators, these relations transfer from $\mathscr{S}$ to the entire L_2, which proves the rest of the proposition. □

Next we study the relation between the L_1-Fourier transformation $u \to \widehat{u}$ given by (2.13) and the L_2-Fourier transformation $u \to Fu$ defined via the described limit process. We start with the following intermediate result:

Lemma 2.2. *The L_2-Fourier transform Fu of a square integrable function u that vanishes outside a bounded set coincides with its L_1-Fourier transform $\widehat{u}$.*

Proof. Since the rapidly decreasing functions form a dense subspace of L_2, there is a sequence $u_1, u_2, \ldots$ of such functions that converge in the L_2-sense to u. We can assume without restriction that the u_k vanish outside a bounded set covering the support of u. As this set has finite measure, the u_k then converge also in the L_1-norm to u and their Fourier transforms $\widehat{u}_k$ hence uniformly to the L_1-Fourier transform of u. The $\widehat{u}_k$ converge on the other hand by definition in the L_2-sense to the L_2-Fourier transform of u. Since uniform convergence implies local L_2-convergence, both limits coincide so that in this case indeed $Fu = \widehat{u}$. □

This observation allows it to determine the L_2-Fourier transform by a limit process that is better suited to explicit calculations and probably also easier to grasp.

Theorem 2.7. *For $u \in L_2$ and $R > 0$ let u_R be the function that attains the same values as u for $|x| \leq R$ and vanishes outside this ball. The L_1-Fourier transforms*

$$\widehat{u}_R(\omega) = \left(\frac{1}{\sqrt{2\pi}}\right)^n \int u_R(x)\, e^{-i\omega \cdot x}\, dx \tag{2.16}$$

of these band-limited functions u_R, uniformly continuous, square integrable functions, tend then in the L_2-sense to the L_2-Fourier transform of u.

Proof. The functions (2.16) are by Lemma 2.2 the L_2-Fourier transforms of the u_R. Since the u_R converge in the L_2-sense to u and the L_2-Fourier transform is a bounded linear operator from L_2 to L_2, the $\widehat{u}_R$ thus converge in the L_2-norm or as one also says in the quadratic mean to the L_2-Fourier transform of u. □

Finally we can consider functions that are contained both in L_1 and L_2. For such functions both kinds of Fourier transformation lead as expected to the same result.

Theorem 2.8. *The L_2-Fourier transform of a both integrable and square integrable function u is its original L_1-Fourier transform given by the integral expression*

$$\widehat{u}(\omega) = \left(\frac{1}{\sqrt{2\pi}}\right)^n \int u(x)\, e^{-i\omega \cdot x}\, dx. \tag{2.17}$$

Proof. The function u_R from the previous theorem converge in this case both in L_1 and L_2 to u and their Fourier transforms therefore uniformly to the L_1-Fourier transform and in the L_2-norm to the L_2-Fourier transform of u. Since uniform convergence implies local L_2-convergence, both limits necessarily coincide. □

We are therefore allowed to denote the L_2-Fourier transform of a square integrable function u without any danger of confusion in the same way as the L_1-Fourier transform of an integrable function u by $\hat{u}$ and will do so from now on.

2.3   Spaces of Weakly Differentiable Functions

The space $\mathscr{S}$ of the rapidly decreasing functions from $\mathbb{R}^n$ to $\mathbb{C}$ or $\mathbb{R}$ and particularly its subspace $\mathscr{D}$ consisting of the functions in $\mathscr{S}$ that vanish outside bounded sets are easy to handle but are much too small for most purposes. In particular they are not complete with respect to the considered norms, that is, Cauchy sequences do not need to converge. As we know, the smallest space that contains the functions in $\mathscr{D}$ and that is complete under the L_2-norm (2.14) is L_2 itself. The space L_2 can therefore be regarded as the completion of $\mathscr{D}$ under the L_2-norm. The aim of this section is to introduce subspaces of L_2 that comprehend $\mathscr{D}$ and $\mathscr{S}$ and are complete under norms measuring also the distance between certain, in an appropriate sense defined partial derivatives. The in the given context most important of these spaces is the space H^1, the completion of $\mathscr{D}$ or $\mathscr{S}$ under the H^1-norm that is given by

$$\|u\|_1^2 = \|u\|_0^2 + |u|_1^2, \quad |u|_1^2 = \|\nabla u\|_0^2, \tag{2.18}$$

and is composed of the L_2-norm of the considered function and the L_2-norm of its first order weak derivatives introduced below.

We begin with the discussion of an approximation process for locally integrable functions that resembles that in the proof of Theorem 2.2. Let $\delta : \mathbb{R}^n \to \mathbb{R}$ be an infinitely differentiable function with values $\delta(x) \geq 0$ that vanishes outside the ball of radius 1 around the origin and has L_1-norm 1. Let $\delta_k(x) = k^n \delta(kx)$ for $k \in \mathbb{N}$. Then

$$\delta_k(x) \geq 0, \quad \delta_k(x) = 0 \text{ for } |x| \geq 1/k, \quad \int \delta_k(x)\,dx = 1. \tag{2.19}$$

For all locally integrable functions u we then define the local averages

$$(\delta_k * u)(x) = \int \delta_k(x-y)\,u(y)\,dy. \tag{2.20}$$

Lemma 2.3. *If u is an integrable function so are its smoothed counterparts given by (2.20). These converge to u in the L_1-sense as k goes to infinity.*

Proof. From (2.19) and Fubini's theorem we get for $u \in L_1$ the estimate

$$\int |(\delta_k * u)(x)| \, dx \leq \iint \delta_k(x-y) |u(y)| \, dy \, dx = \int |u(y)| \, dy$$

for the L_1-norm of $\delta_k * u$. Let $u \in L_1$ now be given and $\tilde{u}$ a function in $\mathscr{D}$ with $\|u - \tilde{u}\|_{L_1} < \varepsilon/4$. The smoothed functions $\delta_k * u$ and $\delta_k * \tilde{u}$ then also differ at most by

$$\|\delta_k * u - \delta_k * \tilde{u}\|_{L_1} = \|\delta_k * (u - \tilde{u})\|_{L_1} \leq \|u - \tilde{u}\|_{L_1} < \varepsilon/4$$

and the error to be estimated can be bounded from above as follows:

$$\|\delta_k * u - u\|_{L_1} < \|\delta_k * \tilde{u} - \tilde{u}\|_{L_1} + \varepsilon/2.$$

Utilizing again the properties (2.19) of the smoothing kernels, we obtain moreover

$$|(\delta_k * \tilde{u})(x) - \tilde{u}(x)| = \left| \int \delta_k(x-y)\{\tilde{u}(y) - \tilde{u}(x)\} \, dy \right| \leq \max_{|x-y| \leq 1/k} |\tilde{u}(x) - \tilde{u}(y)|.$$

Since $\tilde{u}$ is uniformly continuous, the $\delta_k * \tilde{u}$ converge therefore uniformly to $\tilde{u}$. Since the $\delta_k * \tilde{u}$ and $\tilde{u}$ itself vanish outside a fixed bounded set, the uniform convergence implies convergence in the L_1-norm. Hence for sufficiently large indices k

$$\|\delta_k * u - u\|_{L_1} < \varepsilon/2 + \varepsilon/2,$$

which demonstrates that the $\delta_k * u$ tend in the L_1-norm to u as k goes to infinity. $\square$

This result can be generalized to the functions in the spaces L_p for $1 \leq p < \infty$. The proof uses for $p > 1$ the Hölder inequality to bound the functions $\delta_k * u$.

Lemma 2.3 has a local counterpart. Let u be a locally integrable function. Consider a ball of radius R and let $v \in L_1$ coincide with u on the ball of radius $R + 1$ with same center and vanish outside this ball. The functions $\delta_k * v$ tend then in the L_1-norm to v. Since u and v and $\delta_k * u$ and $\delta_k * v$ coincide on the original ball of radius R, the $\delta_k * u$ tend on this ball, and with that on every bounded measurable set, in the L_1-sense to u. A rather immediate consequence of this fact is:

Lemma 2.4. *A locally integrable function u, for which*

$$\int u\varphi \, dx = 0 \tag{2.21}$$

holds for all functions $\varphi \in \mathscr{D}$, *vanishes.*

Proof. The assumption particularly implies that the integrals (2.20) vanish for all k and all x. The proposition follows therefore from the just made observation. $\square$

We remark that the proposition follows for locally square integrable functions more or less directly from the density of the functions $\varphi \in \mathscr{D}$ in L_2 and that in this case one does not need to make a detour via an approximation process as in the more general case of only locally integrable functions. Lemma 2.4 forms the basis of the following generalization of the notion of partial derivative:

Definition 2.3. A locally integrable function $D^\alpha u : \mathbb{R}^n \to \mathbb{C}$, α a multi-index with nonnegative integer components, is denoted as weak partial derivative of corresponding order of the locally integrable function u, if for all test functions $\varphi \in \mathscr{D}$

$$\int D^\alpha u \, \varphi \, dx = (-1)^{|\alpha|} \int u \, D^\alpha \varphi \, dx. \tag{2.22}$$

This definition requires some comments. The first is that the weak derivative $D^\alpha u$ is unique as long as it exists, a fact that first justifies the definition and that follows from Lemma 2.4. The second observation is that sufficiently smooth functions are weakly differentiable in the given sense. Their weak partial derivatives coincide in this case with their normal, classically defined partial derivatives. This is shown with help of Fubini's theorem and integration by parts. The existence of weak derivatives does not however mean that the corresponding classical derivatives must exist.

We recall from Sect. 2.1 that the Fourier transform of the partial derivative $D^\alpha u$, α a multi-index with nonnegative integer components, of a function $u \in \mathscr{S}$ is

$$(\widehat{D^\alpha u})(\omega) = (i\omega)^\alpha \, \widehat{u}(\omega) \tag{2.23}$$

and that, due to Plancherel's theorem, its L_2 norm is given by

$$\|D^\alpha u\|_0^2 = \int \omega^{2\alpha} |\widehat{u}(\omega)|^2 \, d\omega. \tag{2.24}$$

These properties can be used to characterize weak derivatives in the L_2-case.

Theorem 2.9. *A square integrable function u possesses a square integrable weak derivative $D^\alpha u$ if and only if the function*

$$\omega \to (i\omega)^\alpha \, \widehat{u}(\omega) \tag{2.25}$$

is also in L_2. The weak derivative $D^\alpha u$ of u is then the Fourier back-transform of the function (2.25) and its L_2-norm therefore again given by the expression (2.24).

Proof. Let the function (2.25) be square integrable and denote by u_α its Fourier back-transform. In terms of the L_2-inner product and the L_2-Fourier transform F then

$$(u_\alpha, \varphi) = (Fu_\alpha, F\varphi) = ((i\omega)^\alpha Fu, F\varphi) = (-1)^{|\alpha|}(Fu, (i\omega)^\alpha F\varphi)$$
$$= (-1)^{|\alpha|}(Fu, FD^\alpha \varphi) = (-1)^{|\alpha|}(u, D^\alpha \varphi).$$

for all $\varphi \in \mathscr{D}$, from which $u_\alpha = D^\alpha u$ follows. The proof of the opposite direction requires some preparation. Let the weak derivative $D^\alpha u$ of $u \in L_2$ exist and be square integrable. Let φ be a rapidly decreasing function and χ an infinitely differentiable function that takes the values $\chi(x) = 1$ for $|x| \leq 1$ and $\chi(x) = 0$ for $|x| \geq 2$. Let

$$\varphi_R(x) = \chi\left(\frac{x}{R}\right)\varphi(x).$$

As follows from the dominated convergence theorem then

$$\lim_{R\to\infty}(D^\alpha u, \varphi_R) = (D^\alpha u, \varphi), \quad \lim_{R\to\infty}(u, D^\alpha \varphi_R) = (u, D^\alpha \varphi).$$

Provided that both u and $D^\alpha u$ are in L_2, the defining relation (2.22) holds therefore not only for the functions $\varphi \in \mathcal{D}$ but for all $\varphi \in \mathcal{S}$. This implies

$$(FD^\alpha u, F\varphi) = (D^\alpha u, \varphi) = (-1)^{|\alpha|}(u, D^\alpha \varphi) = (-1)^{|\alpha|}(Fu, FD^\alpha \varphi)$$

for all rapidly decreasing functions φ. Since $FD^\alpha \varphi = (i\omega)^\alpha F\varphi$, thus

$$\int FD^\alpha u \overline{F\varphi}\, d\omega = \int (i\omega)^\alpha Fu \overline{F\varphi}\, d\omega$$

for all rapidly decreasing functions φ. As every function $\varphi \in \mathcal{D}$ can itself be written as Fourier transform of a rapidly decreasing function, hence for all functions $\varphi \in \mathcal{D}$

$$\int (i\omega)^\alpha Fu\,\varphi\, d\omega = \int FD^\alpha u\,\varphi\, d\omega.$$

The locally integrable function $\omega \to (i\omega)^\alpha \widehat{u}(\omega)$ and the function $FD^\alpha u \in L_2$ thus coincide by Lemma 2.4 and the first one is as asserted square integrable.   □

Let A be a finite set of multi-indices α with nonnegative integer components that contains the multi-index $\alpha = 0$. To each such set of indices A we assign a subspace H_A of L_2. It consists of the square integrable functions u that possess weak derivatives $D^\alpha u \in L_2$ for all $\alpha \in A$ and is equipped with the norm given by

$$\|u\|_A^2 = \sum_{\alpha \in A}\|D^\alpha u\|_0^2. \tag{2.26}$$

An example of such an index set A is the set of all multi-indices $\alpha = (\alpha_1, \ldots, \alpha_n)$ of order $|\alpha| = \alpha_1 + \ldots + \alpha_n \leq m$. The corresponding space is the Sobolev space H^m that is invariant under rotations. It should however be emphasized that the construction is not restricted to such familiar cases. Another important example is the space of the functions with L_2-bounded m-th order mixed derivatives that corresponds to the set A of the multi-indices α for which $\alpha_i \leq m$ for each component individually.

Theorem 2.10. *The spaces H_A are complete, that is, are Hilbert spaces.*

Proof. Let u_k, $k = 1, 2, \ldots$, be square integrable functions in H_A that form a Cauchy sequence in the sense of the norm given by (2.26). Since L_2 is complete, the functions u_k converge in the L_2-sense to a limit function $u \in L_2$ and their weak derivatives $D^\alpha u_k$, $\alpha \in A$, in the L_2-sense individually to limit functions $v_\alpha \in L_2$. Then

$$(v_\alpha, \varphi) = \lim_{k \to \infty} (D^\alpha u_k, \varphi) = (-1)^{|\alpha|} \lim_{k \to \infty} (u_k, D^\alpha \varphi) = (-1)^{|\alpha|} (u, D^\alpha \varphi)$$

for all test functions $\varphi \in \mathscr{D}$, that is, v_α is the weak derivative $D^\alpha u$ of u.   □

The norm (2.26) on H_A can with help of Theorem 2.9 be written as

$$\|u\|_A^2 = \sum_{\alpha \in A} \int \omega^{2\alpha} |\hat{u}(\omega)|^2 \, d\omega. \tag{2.27}$$

The space H_A can be considered as the completion of $\mathscr{D}$ or $\mathscr{S}$ under this norm:

Theorem 2.11. *The space $\mathscr{D}$ of the infinitely differentiable functions with compact support and with that also the space $\mathscr{S}$ of the rapidly decreasing functions are dense subspaces of all these spaces H_A, independent of the structure of the index sets A.*

Proof. We assign to each function $u \in L_2$ the infinitely differentiable functions

$$u_R(x) = \left(\frac{1}{\sqrt{2\pi}}\right)^n \int_{|\omega| \le R} \hat{u}(\omega) e^{i\omega \cdot x} \, d\omega$$

whose partial derivatives are all square integrable. For $u \in H_A$

$$\|u - u_R\|_A^2 = \sum_{\alpha \in A} \int_{|\omega| > R} \omega^{2\alpha} |\hat{u}(\omega)|^2 \, d\omega,$$

so that the u_R converge for these u in the norm (2.27) to u for R tending to infinity. It suffices therefore to show that every infinitely differentiable function v whose partial derivatives of arbitrary order are square integrable can, in the sense of the norm (2.26), be approximated arbitrarily well by functions in $\mathscr{D}$. For that purpose let χ be as above an infinitely differentiable cut-off function that attains the values $\chi(x) = 1$ for $|x| \le 1$ and $\chi(x) = 0$ for $|x| \ge 2$. The infinitely differentiable functions

$$x \to \chi\left(\frac{x}{R}\right) v(x)$$

vanish then outside the balls of radius $2R$ around the origin and converge in the norm (2.26) to v for R tending to infinity.   □

The fact that the infinitely differentiable functions with compact support are dense in such spaces is of great practical value since it allows to prove many results and estimates first only for these functions and to transfer them then later with the help of continuity arguments to the full space. We will utilize this property often.

For nonnegative integer values m of s the norm on the spaces H^s introduced above is equivalent to the norm given by the expression

$$\|u\|_s^2 = \int \left(1 + |\omega|^{2s}\right)|\hat{u}(\omega)|^2 \, d\omega. \tag{2.28}$$

This expression can be used to define new norms for non-integer values $s \geq 0$ and to introduce corresponding spaces H^s as completions of the spaces $\mathscr{S}$ or $\mathscr{D}$ under this norm. The smoothness of these functions depends on s as follows:

Theorem 2.12. *The functions in H^s are continuous for all indices $s > n/2$ and even m-times continuously differentiable if $s > m + n/2$.*

Proof. We approximate $u \in H^s$ by the infinitely differentiable functions

$$u_k(x) = \left(\frac{1}{\sqrt{2\pi}}\right)^n \int_{|\omega| \leq 2^k} \hat{u}(\omega) e^{i\omega \cdot x} \, d\omega, \quad k = 0, 1, \dots,$$

similarly as above. Their differences satisfy for indices $\ell > k$ the estimate

$$\left|u_\ell(x) - u_k(x)\right|^2 \leq \left(\frac{1}{2\pi}\right)^n \int_{2^k \leq |\omega| \leq 2^\ell} |\omega|^{-2s} \, d\omega \int_{2^k \leq |\omega| \leq 2^\ell} |\omega|^{2s} |\hat{u}(\omega)|^2 \, d\omega.$$

The first of the two integrals on the right hand side takes the value

$$\int_{2^k \leq |\omega| \leq 2^\ell} |\omega|^{-2s} \, d\omega = \sum_{j=k}^{\ell-1} (2^{n-2s})^j \int_{1 \leq |\omega| \leq 2} |\omega|^{-2s} \, d\omega$$

and becomes arbitrarily small for sufficiently big k, provided that $s > n/2$, and the second one can be estimated by the square of the H^s-norm of u. The u_k converge therefore not only in the L_2-sense, but also uniformly so that the limit function is continuous. The same kind of arguments shows that their derivatives up to order m converge uniformly provided that $s > m + n/2$. □

2.4 Fourier and Laplace Transformation

Let $F : \mathbb{R}_{\geq 0} \to \mathbb{C}$ be a measurable function for which there exists a real number s_0 such that the functions $t \to F(t) e^{-st}$ are square integrable over the interval $t \geq 0$ for all $s > s_0$, which is particularly the case if $F(t) e^{-s_0 t}$ is bounded. The function

$$f(z) = \int_0^\infty F(t) e^{-zt} \, dt \tag{2.29}$$

is then defined on the half-plane consisting of all complex numbers z with real part $\mathrm{Re}\, z > s_0$. It is analytic there and is denoted as the Laplace transform of F.

The Laplace and the one-dimensional Fourier transformation are closely related to each other as becomes obvious from the proof of the following inversion theorem:

Theorem 2.13. *Under the given assumptions*

$$F(t)\,e^{-st} = \frac{1}{2\pi} \int_{-\infty}^{\infty} f(s+i\omega)\,e^{i\omega t}\,d\omega \tag{2.30}$$

holds for all real $s > s_0$, *where the function on the right hand side has to be understood as the* L_2-*limit of the infinitely differentiable, square integrable functions*

$$t \longrightarrow \frac{1}{2\pi} \int_{-R}^{R} f(s+i\omega)\,e^{i\omega t}\,d\omega \tag{2.31}$$

for R tending to infinity. The equation (2.30) *has to be interpreted correspondingly in the sense of the equality of functions in* L_2, *that is, as equality almost everywhere.*

Proof. The Laplace transform of F can be represented via the Fourier transforms

$$f(s+i\omega) = \frac{1}{\sqrt{2\pi}} \int_{-\infty}^{\infty} g_s(t)\,e^{-i\omega t}\,dt = \widehat{g}_s(\omega)$$

of the both integrable and square integrable, parameter dependent functions g_s taking the values $g_s(t) = \sqrt{2\pi}\,F(t)\,e^{-st}$ for $t \geq 0$ and $g_s(t) = 0$ for $t < 0$. These functions, and with that also F, can be recovered from their Fourier transforms with help of Theorem 2.6, the L_2-version of the Fourier inversion theorem, and Theorem 2.7. Translating the result into the original notations one obtains (2.30). □

The Laplace inversion formula (2.30) is mostly written as limit

$$F(t) = \lim_{R \to \infty} \frac{1}{2\pi i} \int_{s-iR}^{s+iR} f(z)\,e^{zt}\,dz \tag{2.32}$$

of complex line integrals. If $f(z)$ is a rational function this limit can be calculated with help of the residue calculus. The Laplace transform plays an important role in electrical engineering and can, for example, be used to convert linear differential equations to algebraic equations. In quantum chemistry it serves mainly to simplify the calculation of integrals and to represent functions in terms of Gauss functions. We will come back to the latter point at the very end of this text.

Chapter 3
The Basics of Quantum Mechanics

This chapter gives a short introduction to quantum mechanics starting from de Broglie's and Schrödinger's wave picture. The emphasis is on the mathematical structure of the theory with the aim to form a sound basis for the later study of the electronic Schrödinger equation. The discussion starts in the first two sections with a heuristic derivation of the Schrödinger equation for a single free particle from which, in the third section, the general mathematical framework of quantum mechanics is derived. The fourth section deals with a particular simple quantum-mechanical system, the harmonic oscillator. The harmonic oscillator serves on one hand as an example of a quantum-mechanical system with completely different properties from the free particle and is ideal to exemplify and illustrate the general concepts of quantum theory. On the other hand the explicit knowledge of its solutions will in later chapters help to develop the mathematical theory further. In the fifth section the weak form of the Schrödinger equation is derived and physically motivated. The equivalence of the weak formulation to the classical operator formulation is shown. In later chapters we will exclusively work with the weak form that is basic for the L_2-theory of partial differential equations. The last section is devoted to many-particle systems. The central point here are the symmetry properties of the many-particle wave functions that are not only fundamental for the structure of matter and responsible for many of the strange properties of quantum systems but that will also turn out to be essential for the regularity theory of the electronic Schrödinger equation and for the study of its complexity.

The chapter is tailored to our later needs and can of course not replace the study of basic textbooks in quantum mechanics like [18] or [63]. A standard reference for quantum chemists is [6]. Mathematicians will like [79, 80], not only because of the impressive visualizations and the accompanying software but also because of its mathematical soundness. The historically most important and influential texts are the monographs [25] of Dirac and [64] of von Neumann. The mathematical framework of quantum mechanics presented in this chapter is due to von Neumann.

H. Yserentant, *Regularity and Approximability of Electronic Wave Functions*,
Lecture Notes in Mathematics 2000, DOI 10.1007/978-3-642-12248-4_3,
© Springer-Verlag Berlin Heidelberg 2010

3.1 Waves, Wave Packets, and Wave Equations

Waves are omnipresent in nature. Modern quantum mechanics had it seeds in the early 1920's in de Broglie's insight into the wave-like behavior of electrons that can be directly observed in scattering experiments and that finally leads to the Schrödinger wave equation, the basic equation for our understanding of atoms and molecules. The purpose of this introductory section is to present the general mathematical framework for the description of such wave phenomena.

We first recall the notion of a plane wave, a complex-valued function

$$\mathbb{R}^d \times \mathbb{R} \to \mathbb{C} : (x,t) \to e^{ik \cdot x - i\omega t}, \tag{3.1}$$

with $k \in \mathbb{R}^d$ the wave vector and $\omega \in \mathbb{R}$ the frequency.[1] At a fixed point x in space the plane wave oscillates with the frequency ω and the period $T = 2\pi/\omega$. The quantity $k \cdot x - \omega t$ is called phase. The points x in space for which the phase attains a given value are located on (hyper-)planes orthogonal to k. These planes have the distance $\lambda = 2\pi/|k|$ to each other, which is the spatial wave length, and move with the phase velocity $\omega/|k|$ in the direction of the wave vector. A dispersion relation

$$\omega = \omega(k) \tag{3.2}$$

assigns to each wave vector k a characteristic frequency. Such dispersion relations fix the physics that is described by this kind of waves. Most common is the case

$$\omega = c|k|, \tag{3.3}$$

which arises, for example, in the propagation of light in vacuum and of electromagnetic waves in general. The phase velocity attains in this case the constant value c.

Plane waves are completely delocalized and attain at every point in space the same absolute value 1. They can, however, be superimposed to wave packets

$$\psi(x,t) = \left(\frac{1}{\sqrt{2\pi}}\right)^d \int \widehat{\psi}_0(k) e^{ik \cdot x - i\omega(k)t} \, dk, \tag{3.4}$$

where we preliminarily suppose that $\widehat{\psi}_0$ is a rapidly decreasing function to avoid any mathematical difficulty. The wave packets remain then for any given time t rapidly decreasing functions of the spatial variable x with the spatial Fourier transform

$$\widehat{\psi}(k,t) = e^{P(ik)t} \widehat{\psi}_0(k), \tag{3.5}$$

where we have set $P(ik) = -i\omega(k)$. As this quantity is purely imaginary, for all t

---

[1] We change the notation in this chapter and denote by ω the time frequency, not the argument of the spatial Fourier transforms as before, which will in this chapter be denoted by k according to the conventions in physics. In the forthcoming chapters we will return to the previously used notation.

$$\int |\psi(x,t)|^2 \, dx = \int |\widehat{\psi}_0(k)|^2 \, dk. \tag{3.6}$$

Correspondingly the L_2-norms of the spatial derivatives of ψ remain constant in time. Provided that the absolute value of $P(ik)$ does not increase more rapidly than a polynomial in k for $|k|$ tending to infinity, the wave packets are infinitely differentiable functions of t and have the time derivatives

$$\frac{\partial^m}{\partial t^m} \psi(x,t) = \left(\frac{1}{\sqrt{2\pi}}\right)^d \int P(ik)^m \, \widehat{\psi}(k,t) \, e^{ik \cdot x} \, dk. \tag{3.7}$$

Remembering Theorem 2.4, that is, that a partial derivative corresponds in the Fourier representation to the multiplication with a polynomial in ik, we can formally write this equation as a so-called pseudo-differential equation

$$\frac{\partial^m}{\partial t^m} \psi(x,t) = P(D)^m \, \psi(x,t) \tag{3.8}$$

that becomes a true differential equation if $\xi \to P(\xi)^m$ is a multivariate polynomial. Consider as an example the dispersion relation (3.3). In this case, (3.8) becomes

$$\frac{\partial^2}{\partial t^2} \psi(x,t) = -c^2 \left(\frac{1}{\sqrt{2\pi}}\right)^d \int |k|^2 \, \widehat{\psi}(k,t) \, e^{ik \cdot x} \, dk. \tag{3.9}$$

that is, the classical second-order wave equation

$$\frac{\partial^2 \psi}{\partial t^2} = c^2 \Delta \psi. \tag{3.10}$$

3.2 The Schrödinger Equation for a Free Particle

When de Broglie postulated the wave nature of matter, the problem was to guess the dispersion relation for the matter waves: to guess, as this hypothesis creates a new kind of physics that cannot be deduced from known theories. A good starting point is Einstein's interpretation of the photoelectric effect. When polished metal plates are irradiated by light of sufficiently short wave length they may emit electrons. The magnitude of the electron current is as expected proportional to the intensity of the light source, but their energy surprisingly to the wave length or the frequency of the incoming light. Einstein's explanation, for which he received the Nobel prize, was that light consists of single light quanta with energy and momentum

$$E = \hbar\omega, \quad p = \hbar k \tag{3.11}$$

depending on the frequency ω and the wave vector k. The quantity

$$\hbar = 1.0545716 \cdot 10^{-34} \, \text{kg}\,\text{m}^2\text{s}^{-1} \tag{3.12}$$

is Planck's constant, an incredibly small quantity of the dimension energy × time called action. The relations (3.11) alone are naturally not sufficient to obtain a dispersion relation. To establish a connection between ω and k or E and p, it is an obvious idea to bring additionally the energy-momentum relation

$$E = \sqrt{c^2|p|^2 + m^2c^4} \tag{3.13}$$

of special relativity for a particle of rest mass m into play, where c denotes the speed of light. For particles in rest it turns into the famous formula $E = mc^2$. It yields the desired dispersion relation and with that the second-order wave equation

$$\frac{\partial^2 \psi}{\partial t^2} = c^2 \Delta \psi + \frac{m^2c^4}{\hbar^2} \psi, \tag{3.14}$$

that was later called the Klein-Gordon equation. This is what Schrödinger initially tried. This equation did not meet his expectations, however, and led to the wrong predictions, as it describes another kind of particles (those with spin zero), not electrons. The correct relativistic equation for a single electron (but unfortunately only for a single one) is the Dirac equation that was found a short time later. He therefore fell back to classical physics and replaced (3.13) by the energy-momentum relation

$$E = \frac{1}{2m}|p|^2 \tag{3.15}$$

from Newtonian mechanics. It leads to the dispersion relation

$$\omega = \frac{\hbar}{2m}|k|^2 \tag{3.16}$$

and finally to the wave equation for a non-relativistic free particle of mass m in absence of external forces, the Schrödinger equation

$$i\hbar \frac{\partial \psi}{\partial t} = -\frac{\hbar^2}{2m} \Delta \psi. \tag{3.17}$$

In contrast to the classical wave equation (3.10) and also to the Klein-Gordon equation (3.14) it contains the imaginary unit and is therefore genuinely an equation for complex-valued functions.

The Schrödinger equation (3.17) is of first order in time. Its solutions

$$\psi(x,t) = \left(\frac{1}{\sqrt{2\pi}}\right)^3 \int e^{-i\frac{\hbar}{2m}|k|^2 t} \, \widehat{\psi}_0(k) \, e^{ik\cdot x} \, dk, \tag{3.18}$$

the wave functions, are uniquely determined by their initial state ψ_0. If ψ_0 is a rapidly decreasing function the solution possesses time derivatives of arbitrary order, and all of them are rapidly decreasing functions of the spatial variables. To avoid technicalities, we assume this for the moment. We further recall that

$$\int |\psi(x,t)|^2 \, dx = \int |\widehat{\psi}(k,t)|^2 \, dk. \qquad (3.19)$$

remains constant in time. We assume in the sequel that this value is normalized to 1, which is basic for the statistical interpretation of the wave functions. The quantities $|\psi|^2$ and $|\widehat{\psi}|^2$ can then be interpreted as probability densities. The integrals

$$\int_\Omega |\psi(x,t)|^2 \, dx, \quad \int_{\widehat{\Omega}} |\widehat{\psi}(k,t)|^2 \, dk \qquad (3.20)$$

represent the probabilities to find the particle at time t in the region Ω of the position space, respectively, the region $\widehat{\Omega}$ of the momentum space. The quantity

$$\int \frac{\hbar^2}{2m} |k|^2 |\widehat{\psi}(k,t)|^2 \, dk, \qquad (3.21)$$

is the expectation value of the kinetic energy. With help of the Hamilton operator

$$H = -\frac{\hbar^2}{2m} \Delta, \qquad (3.22)$$

this expectation value can be rewritten as

$$\int \psi \, \overline{H\psi} \, dx = (\psi, H\psi). \qquad (3.23)$$

The expectation values of the components of the momentum are in vector notation

$$\int \hbar k \, |\widehat{\psi}(k,t)|^2 \, dk. \qquad (3.24)$$

Introducing the momentum operator

$$p = -i\hbar\nabla \qquad (3.25)$$

their position representation is the inner product

$$\int \psi \, \overline{p\psi} \, dx = (\psi, p\psi). \qquad (3.26)$$

The expectation values of the three components of the particle position are finally

$$\int x \, |\psi(x,t)|^2 \, dx = (\psi, q\psi), \qquad (3.27)$$

with q the position operator given by $\psi \rightarrow x\psi$. This coincidence between observable physical quantities like energy, momentum, or position and operators acting upon the wave functions is in no way accidental. It forms the heart of quantum mechanics.

3.3 The Mathematical Framework of Quantum Mechanics

We have seen that the physical state of a free particle at a given time t is completely determined by a function in the Hilbert space L_2 that again depends uniquely on the state at a given initial time. In the case of more general systems, the space L_2 is replaced by another Hilbert space, but the general concept remains:

Postulate 1. A quantum-mechanical system consists of a complex Hilbert space $\mathcal{H}$ and a one-parameter group $U(t), t \in \mathbb{R}$, of unitary linear operators on $\mathcal{H}$ with

$$U(0) = I, \quad U(s+t) = U(s)U(t) \tag{3.28}$$

that is strongly continuous in the sense that for all $\psi \in \mathcal{H}$ in the Hilbert space norm

$$\lim_{t \to 0} U(t)\psi = \psi. \tag{3.29}$$

A state of the system corresponds to a normalized vector in $\mathcal{H}$ and the time evolution of the system is described by the group of the propagators $U(t)$. The state

$$\psi(t) = U(t)\psi(0) \tag{3.30}$$

of the system at time t is uniquely determined by its state at time $t = 0$.

In the case of free particles considered so far, the solution of the Schrödinger equation and with that time evolution is given by (3.18). The evolution operators $U(t)$, or propagators, read therefore in the Fourier or momentum representation

$$\widehat{\psi}(k) \rightarrow e^{-i\frac{\hbar}{2m}|k|^2 t} \widehat{\psi}(k). \tag{3.31}$$

Strictly speaking, they have first only been defined for rapidly decreasing functions, functions in a dense subspace of L_2, but it is obvious from Plancherel's theorem that they can be uniquely extended from there to L_2 and have the required properties.

The next step is to move from Postulate 1 to an abstract version of the Schrödinger equation. For that we have to establish a connection between such strongly continuous groups of unitary operators and abstract Hamilton operators.

Definition 3.1. Let $D(H)$ be the linear subspace of the given system Hilbert space $\mathcal{H}$ that consists of those elements ψ in $\mathcal{H}$ for which the limit

$$H\psi = i\hbar \lim_{\tau \to 0} \frac{U(\tau) - I}{\tau} \psi \tag{3.32}$$

exists in the sense of norm convergence. The mapping $\psi \to H\psi$ from the domain $D(H)$ into the Hilbert space $\mathcal{H}$ is then called the generator H of the group.

To determine the generator for the case of the free particle, that is, for the unitary operators $U(t)$ from L_2 to L_2 given by (3.31), we first calculate the expression

$$\left\| i\hbar \, \frac{U(\tau) - I}{\tau} \, \psi + \frac{\hbar^2}{2m} \, \Delta\psi \right\|_0^2 \tag{3.33}$$

for functions $\psi \in H^2$. Setting $\vartheta = \hbar\tau/2m$, its Fourier representation reads

$$\left(\frac{\hbar^2}{2m}\right)^2 \int \left| i \, \frac{e^{-i|k|^2\vartheta} - 1}{|k|^2\vartheta} - 1 \right|^2 |k|^4 |\widehat{\psi}(k)|^2 \, dk. \tag{3.34}$$

The norm (3.33) tends therefore to zero as $\tau \to 0$ by the dominated convergence theorem. For every $\psi \in L_2$ for which the limit (3.32) exists and every $R > 0$ conversely

$$\|H\psi\|_0^2 \geq \left(\frac{\hbar^2}{2m}\right)^2 \int_{|k|\leq R} |k|^4 |\widehat{\psi}(k)|^2 \, dk, \tag{3.35}$$

so that the H^2-norm of such a function ψ must remain finite.[2] The generator of the evolution operator of the free particle is therefore the operator

$$H = -\frac{\hbar^2}{2m} \, \Delta \tag{3.36}$$

with the Sobolev space H^2 as domain of definition $D(H)$. In view of this observation the following result for the general abstract case is unsurprising:

Theorem 3.1. *For all initial values $\psi(0)$ in the domain $D(H)$ of the generator of the group of the propagators $U(t)$, the elements (3.30) are contained in $D(H)$, too, depend continuously differentiable on t, and satisfy the differential equation*

$$i\hbar \, \frac{d}{dt} \, \psi(t) = H\psi(t). \tag{3.37}$$

Proof. For all elements $\psi(0)$ in $D(H)$ and all t, the limit

$$\lim_{\tau \to 0} \frac{U(\tau) - I}{\tau} \, U(t)\psi(0) = \lim_{\tau \to 0} U(t) \, \frac{U(\tau) - I}{\tau} \, \psi(0) = -\frac{i}{\hbar} \, U(t)H\psi(0)$$

exists, which means that $\psi(t) = U(t)\psi(0)$ is contained in $D(H)$. Therefore

$$i\hbar \, \lim_{\tau \to 0} \frac{\psi(t + \tau) - \psi(t)}{\tau} = i\hbar \, \lim_{\tau \to 0} \frac{U(\tau) - I}{\tau} \, \psi(t) = H\psi(t),$$

which shows that $t \to \psi(t)$ is a strong solution of (3.37), whose derivative

$$\psi'(t) = -\frac{i}{\hbar} \, U(t)H\psi(0)$$

depends because of the strong continuity of the group continuously on t. □

---

[2] Unfortunately, the Sobolev spaces $H^1, H^2, \ldots$ are denoted by the same letter as the generator H of the group, the Hamiltonian of the system. Both notations are common, so we keep them here.

It should noted once more, however, that the differential equation (3.37), the abstract Schrödinger equation, makes sense only for initial values in the domain of the generator H, but that the propagators are defined on the whole Hilbert space.

A little calculation shows that for the solutions $\psi, \phi : \mathbb{R} \to D(H)$ of (3.37)

$$0 = i\hbar \frac{\mathrm{d}}{\mathrm{d}t} (\psi(t), \phi(t)) = (H\psi(t), \phi(t)) - (\psi(t), H\phi(t)). \qquad (3.38)$$

For all ψ and ϕ in the domain $D(H)$ of the generator H therefore

$$(H\psi, \phi) = (\psi, H\phi). \qquad (3.39)$$

The generators of one-parameter unitary groups are thus necessarily symmetric. Symmetry alone does, however, not suffice to characterize them completely.

Definition 3.2. Let $A : D(A) \to \mathscr{H}$ be a linear operator that is defined on a dense subspace $D(A)$ of $\mathscr{H}$. Let $D(A^\dagger)$ be the set of all $\phi \in \mathscr{H}$ for which there exists an element $\xi \in \mathscr{H}$ with $(\xi, \psi) = (\phi, A\psi)$ for all $\psi \in D(A)$. As $D(A)$ is dense in $\mathscr{H}$ this ξ is then also uniquely determined, so that one can define by $A^\dagger \phi = \xi$ a new mapping $A^\dagger$ from $D(A^\dagger)$ to $\mathscr{H}$, called the adjoint of A. The operator A is called self-adjoint if $A^\dagger = A$ and in particular the domains $D(A^\dagger)$ and $D(A)$ coincide.

This is a very subtle definition. Self-adjointness is more than symmetry. Symmetry only means that $A^\dagger$ is an extension of A to a possibly larger domain $D(A^\dagger)$, self-adjointness that the domain of A is in some sense already maximal. The Hamilton operator (3.36) is an example of a self-adjoint operator with the Sobolev space H^2 as domain of definition. This can be easily proved with help of the Fourier representation and is no accidental coincidence, as follows from the next theorem, Stone's theorem, a cornerstone in the mathematical foundation of quantum mechanics:

Theorem 3.2. *If $U(t), t \in \mathbb{R}$, is a one-parameter unitary group as in Postulate 1, the domain $D(H)$ of its generator H is a dense subset of the underlying Hilbert space and the generator itself self-adjoint. Every self-adjoint operator H is conversely the generator of such a one-parameter unitary group, that is usually denoted as*

$$U(t) = \mathrm{e}^{-\frac{i}{\hbar} Ht}. \qquad (3.40)$$

Proof. Since we are primarily interested in stationary states and will not further refer to Stone's theorem we give only a short sketch of the proof. For some special cases the more important second part of the theorem is easily shown. When H is bounded one defines the evolution operators (3.40) simply with help of the power series expansion of the exponential function. If H possesses a complete set of eigenvectors $\psi_1, \psi_2, \ldots$, the evolution operator can be written down in terms of the corresponding eigenvector expansion of the vector to which it is applied. The proof for the general case is correspondingly based on the spectral decomposition

$$A = \int_{-\infty}^{\infty} \lambda \, dE_\lambda$$

of self-adjoint operators that von Neumann [64] developed to establish a sound mathematical basis for quantum mechanics which was quite new at the time, essentially the framework described here. The unitary group that a self-adjoint operator generates can be easily given in terms of its spectral decomposition and reads

$$e^{-iAt} = \int_{-\infty}^{\infty} e^{-i\lambda t} dE_\lambda.$$

Details can be found in textbooks and monographs on functional analysis, like [69, 87], or [91]. The reverse direction, that the generator of such a unitary group is a self-adjoint operator, can be proven by more elementary means; see [87]. □

Instead of the unitary group of the propagators, a quantum-mechanical system can be thus equivalently fixed by the generator H of this group, the Hamilton operator, or in the language of physics, the Hamiltonian of the system.

In our discussion of the free particle we have seen that there is a direct correspondence between the expectation values of the energy, the momentum, and the position of the particle and the energy or Hamilton operator (3.22), (3.36), the momentum operator (3.25), and the position operator $x \to x\psi$. Each of these operators is self-adjoint. The Hamilton operator has already been discussed, its domain is the Sobolev space H^2. For the momentum operator this is seen by means of its Fourier representation; its domain is the Sobolev space H^1. The domain of the position operator consists of all those wave functions ψ for which $x \to x\psi$ is still square integrable. This reflects the general structure of quantum mechanics:

Postulate 2. Observable physical quantities, or observables, are in quantum mechanics represented by self-adjoint operators $A : D(A) \to \mathcal{H}$ defined on dense subspaces $D(A)$ of the system Hilbert space $\mathcal{H}$. The quantity

$$\langle A \rangle = (\psi, A\psi) \tag{3.41}$$

is the expectation value of a measurement of A for the system in state $\psi \in D(A)$.

At this point we have to recall the statistical nature of quantum mechanics. Quantum mechanics does not make predictions on the outcome of a single measurement of a quantity A but only on the mean result of a large number of measurements on "identically prepared" states, that is, on a given $\psi \in D(A)$. The quantity (3.41) has thus to be interpreted as the mean result that one obtains from a large number of such measurements. This gives reason to consider the standard deviation or uncertainty

$$\Delta A = \|A\psi - \langle A \rangle \psi\|. \tag{3.42}$$

The uncertainty is zero if and only if $A\psi = \langle A \rangle \psi$, that is, if ψ is an eigenvector of A for the eigenvalue $\lambda = \langle A \rangle$. Only in such eigenstates the quantity represented by the operator A can be sharply measured without uncertainty.

One of the fundamental results of quantum mechanics is that, only in exceptional cases, can different physical quantities be measured simultaneously without uncertainty, the Heisenberg uncertainty principle. Its abstract version reads as follows:

Theorem 3.3. *Let A and B two self-adjoint operators and let ψ be a normalized state in the intersection of $D(A)$ and $D(B)$ such that $A\psi \in D(B)$ and $B\psi \in D(A)$. The product of the corresponding uncertainties is then bounded from below by*

$$\Delta A\, \Delta B \geq \frac{1}{2}|((BA - AB)\psi, \psi)|. \tag{3.43}$$

Proof. For arbitrarily chosen real values λ and μ,

$$((BA - AB)\psi, \psi) = ((B - \mu I)(A - \lambda I)\psi, \psi) - ((A - \lambda I)(B - \mu I)\psi, \psi).$$

Since the operators A and B are self-adjoint, one can rearrange this to

$$((BA - AB)\psi, \psi) = 2\mathrm{i}\,\mathrm{Im}((A - \lambda I)\psi, (B - \mu I)\psi).$$

The Cauchy-Schwarz inequality yields

$$|((BA - AB)\psi, \psi)| \leq 2\,\|A\psi - \lambda\,\psi\|\,\|B\psi - \mu\psi\|.$$

The expression on the right hand side attains its minimum if one inserts the expectation values $\langle A \rangle = (\psi, A\psi)$ and $\langle B \rangle = (\psi, B\psi)$ for λ and μ. This proves (3.43). $\qquad\square$

As an example we consider the three components

$$q_k = x_k, \quad p_k = -\mathrm{i}\hbar\frac{\partial}{\partial x_k} \tag{3.44}$$

of the position and the momentum operator. Their commutators are

$$q_k p_k - p_k q_k = \mathrm{i}\hbar I. \tag{3.45}$$

This results in the Heisenberg uncertainty principle

$$\Delta p_k\, \Delta q_k \geq \frac{1}{2}\hbar. \tag{3.46}$$

Position and momentum therefore can never be determined simultaneously without uncertainty, independent of the considered state of the system. The inequality (3.46) and with that also (3.43) are sharp as the instructive example

$$\psi(x) = \left(\frac{1}{\sqrt{\vartheta}}\right)^3 \psi_0\left(\frac{x}{\vartheta}\right), \quad \psi_0(x) = \left(\frac{1}{\sqrt{\pi}}\right)^{3/2}\exp\left(-\frac{1}{2}|x|^2\right), \tag{3.47}$$

of three-dimensional Gauss functions of arbitrary width demonstrates. For these wave functions the inequality (3.46) actually turns into an equality. From

$$\widehat{\psi}(k) = (\sqrt{\vartheta})^3 \, \psi_0(\vartheta k) \tag{3.48}$$

one recognizes that a sharp localization in space, that is, a small parameter ϑ determining the width of ψ, is combined with a loss of localization in momentum.

States with a well defined, sharp energy E play a particularly important role in quantum mechanics, that is, solutions $\psi \neq 0$ in $\mathscr{H}$ of the eigenvalue problem

$$H\psi = E\psi, \tag{3.49}$$

the stationary Schrödinger equation. The functions

$$t \to e^{-i\frac{E}{\hbar}t}\psi \tag{3.50}$$

represent then solutions of the original time-dependent Schrödinger equation. Our main focus in the forthcoming chapters will be on stationary Schrödinger equations.

3.4 The Harmonic Oscillator and Its Eigenfunctions

The Hamilton operator (3.36) of the free particle has no eigenfunction in its domain, the Sobolev space H^2, as can be shown switching to the Fourier representation. This behavior differs completely from that of the system considered in this section, the harmonic oscillator. The harmonic oscillator is one of the few quantum-mechanical systems for which the Schrödinger equation can be solved exactly. It can serve to describe the behavior of quantum-mechanical systems in the neighborhood of points of equilibrium and plays therefore in quantum theory at least as important a role as its classical counterpart in the description of macroscopic systems. The system Hilbert space of the harmonic oscillator is again the space L_2 of the square integrable functions. To find its Hamiltonian, we start from the Hamilton function

$$H(p,q) = \frac{1}{2m}|p|^2 + \frac{m\omega^2}{2}|q|^2 \tag{3.51}$$

of classical mechanics, where m denotes the mass of the considered particle, p is its momentum, q its position and ω the oscillator frequency. The first part on the right hand side represents the kinetic energy and the second the potential energy

$$V(q) = \frac{m\omega^2}{2}|q|^2 \tag{3.52}$$

as a function of the position. The correspondence principle, a collection of rules describing the transition from classical mechanics to quantum mechanics, tells us that the Hamilton operator of the harmonic oscillator reads therefore

$$H = -\frac{\hbar^2}{2m}\Delta\psi + V(x)\psi, \tag{3.53}$$

with V the potential (3.52) just introduced representing the potential energy and the Laplacian part the kinetic energy as for the free particle. The domain of this operator contains the rapidly decreasing functions. The Schrödinger equation (3.37) of the quantum-mechanical harmonic oscillator is thus given by

$$i\hbar \frac{\partial \psi}{\partial t} = -\frac{\hbar^2}{2m} \Delta \psi + \frac{m\omega^2}{2} |x|^2 \psi, \tag{3.54}$$

and its stationary counterpart counterpart (3.49) finally becomes

$$-\frac{\hbar^2}{2m} \Delta \psi + \frac{m\omega^2}{2} |x|^2 \psi = E\psi. \tag{3.55}$$

The solutions of these equations for different values of the mass and the oscillator frequency transfer to each other by scaling. In terms of the dimensionless quantities

$$x' = \frac{x}{L}, \quad t' = \frac{t}{T}; \quad L = \sqrt{\frac{\hbar}{m\omega}}, \quad T = \frac{1}{\omega}, \tag{3.56}$$

omitting the dashes the time-dependent Schrödinger (3.54) equation reads

$$i\frac{\partial \psi}{\partial t} = -\frac{1}{2} \Delta \psi + \frac{1}{2} |x|^2 \psi, \tag{3.57}$$

and with the rescaled energies $\lambda = E/\hbar\omega$, its stationary counterpart (3.55) becomes

$$-\frac{1}{2} \Delta \psi + \frac{1}{2} |x|^2 \psi = \lambda \psi. \tag{3.58}$$

Our next aim is to study the solutions of this eigenvalue problem. As the Hamilton operator splits into a sum of operators each acting only on a single component, we can restrict ourselves essentially to the one-dimensional case, that is, to the operator

$$H = -\frac{1}{2} \frac{d^2}{dx^2} + \frac{1}{2} x^2. \tag{3.59}$$

Surprisingly, this eigenvalue problem can be solved almost without any computation. The crucial observation, due to Dirac, is that this operator can be written as

$$H = A^\dagger A + \frac{1}{2}, \tag{3.60}$$

with A and $A^\dagger$ the two formally adjoint first order "ladder" operators

$$A = \frac{1}{\sqrt{2}} \left(\frac{d}{dx} + x \right), \quad A^\dagger = \frac{1}{\sqrt{2}} \left(-\frac{d}{dx} + x \right), \tag{3.61}$$

whose name becomes obvious from the following considerations.

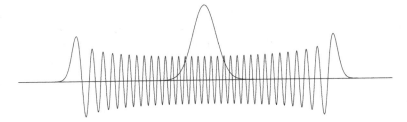

Fig. 3.1 The eigenfunctions ϕ_0 and ϕ_{64} of the one-dimensional harmonic oscillator

Theorem 3.4. *The Hamilton operator (3.59) has the normalized eigenfunctions*

$$\phi_0(x) = \left(\frac{1}{\pi}\right)^{1/4} e^{-x^2/2}, \quad \phi_n = \frac{1}{\sqrt{n!}} (A^\dagger)^n \phi_0 \tag{3.62}$$

that are polynomial multiples of the Gauss function ϕ_0, with assigned eigenvalues

$$\lambda_n = n + \frac{1}{2}, \quad n = 0, 1, 2, \ldots . \tag{3.63}$$

Proof. It suffices to show that the ϕ_n are eigenfunctions of $A^\dagger A$. The essential property of the operators A and $A^\dagger$ and the key to the proof is the commutation relation

$$AA^\dagger = A^\dagger A + 1. \tag{3.64}$$

Since $A\phi_0 = 0$ and $(A^\dagger v, w) = (v, Aw)$ for all rapidly decreasing functions v and w,

$$A^\dagger A \phi_n = n \phi_n, \quad (\phi_n, \phi_n) = 1,$$

as follows by simple induction. This already completes the proof. $\qquad\square$

The operator $A^\dagger$ increases the excitation level by one, it climbs up the ladder. Its counterpart A steps down and decreases the excitation level by one. In formulas:

$$\phi_{n+1} = \frac{1}{\sqrt{n+1}} A^\dagger \phi_n, \quad \phi_{n-1} = \frac{1}{\sqrt{n}} A \phi_n. \tag{3.65}$$

The next question is whether there are further eigenfunctions and eigenvalues, or the other way around, whether the finite linear combinations of the functions (3.62) form a dense subset of L_2. The answer to this question requires some preparations. We start from the following variant of Wiener's density theorem:

Lemma 3.1. *The finite linear combinations of the shifted Gauss functions*

$$g_a(x) = \exp\left(-\frac{(x-a)^2}{2}\right), \quad a \in \mathbb{R}, \tag{3.66}$$

form a dense subset of the space L_2 of the square integrable functions.

Proof. The Hilbert space L_2 can be decomposed into the direct sum of the closure of the linear combinations of the functions (3.66) and the orthogonal complement of this subspace. It suffices therefore to prove that this orthogonal complement consists only of the function $f = 0$, that is, that $(f, g_a) = 0$ for all $a \in \mathbb{R}$ implies that the function f vanishes. The Fourier transforms of the g_a are by Lemma 2.1

$$\widehat{g}_a(k) = e^{-ika} g_0(k).$$

Plancherel's theorem yields therefore, for $f \in L_2$ arbitrary,

$$(f, g_a) = \int_{-\infty}^{\infty} \widehat{f}(k) g_0(k) e^{ika} \, dk.$$

Since $f \in L_2$ implies $\widehat{f} \in L_2$ and with that $\widehat{f} g_0 \in L_1 \cap L_2$, the orthogonality of $f \in L_2$ to all functions g_a thus implies that the Fourier transform of $\widehat{f} g_0$ and with that $\widehat{f} g_0$ itself vanish. As $g_0(k) \neq 0$ for all k, this means $\widehat{f} = 0$ and finally $f = 0$.   □

We remark that result can be generalized; the arguments transfer to the translates of every function in L_2 with bounded, strictly positive Fourier transform.

Lemma 3.2. *The shifted Gauss function (3.66) can be approximated arbitrarily well in the L_2-sense by finite linear combinations of the eigenfunctions (3.62).*

Proof. Inserting the power series expansion of $x \to e^{ax}$, one gets the representation

$$g_a(x) = e^{-a^2/2} \sum_{n=0}^{\infty} \frac{a^n}{n!} x^n e^{-x^2/2}.$$

This series converges not only locally uniformly but also in the L_2-sense, since the sum of the L_2-norms of the single summands remains finite and L_2 is complete. As

$$\int_{-\infty}^{\infty} x^{2n} e^{-x^2} \, dx = \frac{2n-1}{2} \int_{-\infty}^{\infty} x^{2n-2} e^{-x^2} \, ds \leq n! \int_{-\infty}^{\infty} e^{-x^2} \, dx$$

for $n \geq 1$, the squares of these norms satisfy namely the estimate

$$\int_{-\infty}^{\infty} \left| \frac{a^n}{n!} x^n e^{-x^2/2} \right|^2 \, dx \leq \frac{a^{2n}}{n!} \int_{-\infty}^{\infty} e^{-x^2} \, dx.$$

As the eigenfunction ϕ_n is the product of the function $x \to e^{-x^2/2}$ with a polynomial of order n with non-vanishing leading coefficient, the single summands in the series can be written as finite linear combinations of the eigenfunctions (3.62).   □

From Lemma 3.1 and Lemma 3.2 we can conclude that the process described in Theorem 3.4 indeed yields all eigenfunctions of the Hamilton operator (3.59):

Theorem 3.5. *The set of the eigenfunctions (3.62) is complete. For all $f \in L_2$*

$$\lim_{N \to \infty} \left\| f - \sum_{n=0}^{N} (f, \phi_n) \, \phi_n \right\|_0 = 0. \tag{3.67}$$

Proof. The two lemmata show that the finite linear combinations of the eigenfunctions are dense in L_2. Since the ϕ_n form an orthonormal system, being eigenfunctions of a self-adjoint operator corresponding to distinct eigenvalues, the projection

$$P_N f = \sum_{n=0}^{N} (f, \phi_n) \, \phi_n \tag{3.68}$$

is the best approximation of $f \in L_2$ by a linear combination of $\phi_1, \phi_2, \ldots, \phi_N$.   □

Particularly further eigenfunctions $\phi \neq 0$ that are orthogonal to all the eigenfunctions ϕ_n from (3.62) cannot exist. Another consequence of (3.67) is the relation

$$\|f\|_0^2 = \sum_{n=0}^{\infty} |(f, \phi_n)|^2 \tag{3.69}$$

between the L_2-norm of a square integrable function f and the ℓ_2-norm of the sequence of its expansion coefficients that is often denoted as Parseval identity.

Next we want to measure and characterize the speed of convergence of the eigenfunction expansion. For that purpose we introduce a scale of norms given by

$$\|f\|_s^2 = \sum_{n=0}^{\infty} (n+1)^{2s} |(f, \phi_n)|^2 \tag{3.70}$$

for $s \geq 0$ arbitrary. These norms should not be confused with the norms on the Sobolev spaces H^s from Sect. 2.3 and have a very direct interpretation for integer values of s in terms of the smoothness and the decay rate of the considered functions:

Lemma 3.3. *For rapidly decreasing functions f and integer values s, the norm given by the expression (3.70) is equivalent to the L_2-norm of the functions $H^s f$.*

Proof. The central observation is that with f also the functions $H^s f$ are rapidly decreasing. From the representation (3.69) of the L_2-norm, from the fact that the operator H is self-adjoint, and the fact that $H\phi_n = \lambda_n \phi_n$ one obtains therefore

$$\|H^s f\|_0^2 = \sum_{n=0}^{\infty} \lambda_n^{2s} |(f, \phi_n)|^2$$

for all nonnegative integers s. Since $\lambda_n = n + 1/2$, this proves the proposition.   □

The lemma shows particularly that the norms (3.70) remain finite for the rapidly decreasing functions. For all r less than s, one gets the error estimate

$$\|f - P_N f\|_r \leq N^{r-s} \|f\|_s. \tag{3.71}$$

Hence the approximation error tends faster to zero than any given power of $1/N$ for functions that are sufficiently smooth and decay rapidly enough. The approximation is not saturated, as with Fourier series. Compared to Fourier series the convergence rate halves with given order of differentiability, a fact that is owed to the infinite extension of the real axis. With help of the usual techniques from approximation theory one can link the convergence rate directly to the given kind of regularity.

If one expresses the wave functions in terms of their eigenfunction expansions, everything reduces to a very simple diagonal form. The Hamilton operator (3.59) itself reads in terms of the eigenfunction expansion

$$H\psi = \sum_{n=0}^{\infty} \lambda_n (\psi, \phi_n) \, \phi_n \tag{3.72}$$

Its domain consists of all those square integrable functions ψ for which the series

$$\sum_{n=0}^{\infty} \lambda_n^2 |(\psi, \phi_n)|^2 \tag{3.73}$$

converges. Remarkably it is much smaller than the subspace of L_2 that consists of the functions ψ with finite energy expectation value, for which the quadratic form

$$(\psi, H\psi) = \sum_{n=0}^{\infty} \lambda_n |(\psi, \phi_n)|^2 \tag{3.74}$$

attains a finite value. This quadratic form induces a norm that is equivalent to the norm given by (3.70) for $s = 1/2$. We will come back to this important observation in the next section. The uncertainty ΔE of a measurement of the energy in the normalized state ψ in the domain of the Hamilton operator H is given by

$$(\Delta E)^2 = \sum_{n=0}^{\infty} (\lambda_n - (\psi, H\psi))^2 |(\psi, \phi_n)|^2. \tag{3.75}$$

The likelihood that the measurement returns a value $\alpha < \lambda \leq \beta$ is

$$\sum_{\alpha < \lambda_n \leq \beta} |(\psi, \phi_n)|^2, \tag{3.76}$$

and the probability that it yields a value outside the spectrum, that is, no eigenvalue, is therefore zero. The unitary group that H generates in the given case is simply

$$U(t)\psi = \sum_{n=0}^{\infty} e^{i\lambda_n t} (\psi, \phi_n) \, \phi_n. \tag{3.77}$$

It is instructive at this place to return for a moment to physical units. If we insert for the mass m of the particle in the Schrödinger equation (3.54) the mass

$$m = 9.1093822 \cdot 10^{-31} \, \text{kg} \tag{3.78}$$

of the electron and choose the frequency ω so that the ground state energy $E = \hbar\omega/2$ of the oscillator coincides with the binding energy of the electron in the hydrogen atom, the characteristic length and the characteristic time in (3.56) become

$$L = 5.2917721 \cdot 10^{-11} \, \text{m}, \quad T = 2.4188843 \cdot 10^{-17} \, \text{s} \tag{3.79}$$

in meters and seconds. The constant L is the atomic length unit, the Bohr, and the constant T the atomic time unit. The atomic energy unit, the Hartree, is $E = \hbar\omega$, or

$$E = 4.3597439 \cdot 10^{-18} \, \text{kg}\,\text{m}^2\,\text{s}^{-2}. \tag{3.80}$$

The Planck constant itself attains in these units the value $\hbar = 1$. These numbers give an impression of the dimensions of the objects that quantum mechanics studies.

The eigenvalue problem for the two-, three- or higher-dimensional case can be easily reduced to the case of one space dimension; all our considerations directly transfer. Since the three-dimensional operator (3.58) splits into the sum

$$-\frac{1}{2}\Delta + \frac{1}{2}|x|^2 = \sum_{i=1}^{3}\left\{ -\frac{1}{2}\frac{\partial^2}{\partial x_i^2} + \frac{1}{2}x_i^2 \right\} \tag{3.81}$$

of three one-dimensional operators (3.59) each of which acts only on one of the three components of x, the eigenfunctions are simply the tensor products

$$\psi(x) = \phi_{n_1}(x_1)\phi_{n_2}(x_2)\phi_{n_3}(x_3), \quad n_1, n_2, n_3 = 0, 1, 2, \dots. \tag{3.82}$$

In contrast to the one-dimensional case the eigenvalues

$$\lambda = n + \frac{3}{2}, \quad n = 0, 1, 2, \dots, \tag{3.83}$$

are highly degenerate. The dimension

$$\frac{(n+1)(n+2)}{2} \tag{3.84}$$

of the corresponding eigenspaces is equal to the number of possibilities to write the nonnegative integer n as a sum $n = n_1 + n_2 + n_3$ of three other nonnegative integers. The eigenfunctions (3.82) span the linear space of the products

$$x \to P(x)\,e^{-|x|^2/2} \tag{3.85}$$

of polynomials P in three variables with a fixed rotationally symmetric Gaussian, a class of functions that is therefore dense in L_2 and is itself invariant under rotations.

3.5 The Weak Form of the Schrödinger Equation

In our discussion of the harmonic oscillator we started from an expression for the total energy of the system which led us to its Hamilton operator H. This approach can be generalized and offers at the same time an elegant possibility to escape from often very serious mathematical difficulties dealing with self-adjoint extensions of Hamilton operators that are in the beginning only defined on much too small spaces of smooth functions. The approach starts from a subspace $\mathscr{H}_1$ of the system Hilbert space $\mathscr{H}$ that is dense in $\mathscr{H}$ and is itself a Hilbert space under a norm $\|\cdot\|_1$ that dominates the given norm $\|\cdot\|$ on $\mathscr{H}$. This space is associated with the elements $\psi \in \mathscr{H}$ with finite expectation value $B(\psi, \psi)$ of the total energy, where

$$B : \mathscr{H}_1 \times \mathscr{H}_1 \rightarrow \mathbb{C} : (\psi, \phi) \rightarrow B(\psi, \phi) \tag{3.86}$$

is a hermitian bounded bilinear form on $\mathscr{H}_1$, a hermitian bilinear form for which

$$|B(\psi, \phi)| \leq M \|\psi\|_1 \|\phi\|_1 \tag{3.87}$$

holds for all elements $\phi, \psi \in \mathscr{H}_1$. Moreover we assume that for all $\psi \in \mathscr{H}_1$

$$B(\psi, \psi) \geq \delta \|\psi\|_1^2 - \mu \|\psi\|^2, \tag{3.88}$$

with δ a positive and μ an arbitrary real constant. In both cases considered so far, in the case of the free particle and of the harmonic oscillator, this bilinear form reads

$$B(\psi, \phi) = (H\psi, \phi), \tag{3.89}$$

for rapidly decreasing wave function ϕ and ψ and can be extended to a much larger Hilbert space. In the case of the free particle, this is the Sobolev space H^1, the space of the square integrable functions for which the expectation value of the kinetic energy remains finite, and in the case of the harmonic oscillator a subspace of H^1. The key observation is that every such bilinear form induces conversely a self-adjoint operator H that is then the Hamilton operator of the system and can in cases as just given be considered as self-adjoint extension of the original differential operator. This is the famous Friedrichs extension that can be summarized as follows:

Theorem 3.6. *The set $D(H)$ of all $\psi \in \mathscr{H}_1$ for which there exists an element $\xi \in \mathscr{H}$ with $B(\psi, \phi) = (\xi, \phi)$ for all $\phi \in \mathscr{H}_1$ forms a dense subspace of $\mathscr{H}_1$ and with that also of $\mathscr{H}$. There is a unique self-adjoint operator $H : D(H) \rightarrow \mathscr{H}$ with*

$$B(\psi, \phi) = (H\psi, \phi) \tag{3.90}$$

for all elements $\psi \in D(H)$ and all elements $\phi \in \mathcal{H}_1$.

Proof. Under the given assumptions, the expression

$$\langle \psi, \phi \rangle = B(\psi, \phi) + \mu(\psi, \phi)$$

defines an inner product on $\mathcal{H}_1$ which induces a norm on $\mathcal{H}_1$ that is equivalent to the original norm and under which $\mathcal{H}_1$ is complete. The Riesz representation theorem thus guarantees that for every $\xi \in \mathcal{H}$ there is a unique $G\xi \in \mathcal{H}_1$ with

$$\langle G\xi, \phi \rangle = (\xi, \phi), \quad \phi \in \mathcal{H}_1.$$

The mapping $G : \mathcal{H} \to \mathcal{H} : \xi \to G\xi$ is linear, bounded, symmetric, and injective. As $B(\psi, \phi) = (\xi, \phi)$ for all $\phi \in \mathcal{H}_1$ if and only if $\psi = G(\xi + \mu\psi)$ and as conversely $B(G\xi, \phi) = (\xi - \mu G\xi, \phi)$ for all $\phi \in \mathcal{H}_1$, the range of G is the set $D(H)$ introduced above. It is a dense subset of $\mathcal{H}_1$ and with that also of $\mathcal{H}$. Let $H_0 : D(H) \to \mathcal{H}$ be the inverse of G and set $H = H_0 - \mu I$. For all $\psi \in D(H)$ and $\phi \in \mathcal{H}_1$ then

$$(H\psi, \phi) = B(\psi, \phi).$$

To calculate the adjoint of H and its domain, let $\phi \in \mathcal{H}$ and $\xi \in \mathcal{H}$ be given. Then

$$(\xi, \psi) = (\phi, H\psi)$$

for all $\psi \in D(H)$, or $(\xi, G\chi) = (\phi, HG\chi) = (\phi, \chi - \mu G\chi)$ for all $\chi \in \mathcal{H}$, if and only if $G\xi = \phi - \mu G\phi$ or $\phi \in D(H)$ and $\xi = H\phi$. This shows that H is self-adjoint. The uniqueness of H follows simply from the density of $\mathcal{H}_1$ in $\mathcal{H}$. □

Let us now consider the stationary Schrödinger equation (3.49), the problem to find the solutions $\psi \neq 0$ in $D(H)$ of the eigenvalue equation

$$H\psi = E\psi. \tag{3.91}$$

By (3.90) a solution $\psi \in D(H)$ of this equation also solves the equation

$$B(\psi, \phi) = E(\psi, \phi), \quad \phi \in \mathcal{H}_1. \tag{3.92}$$

If conversely $\psi \in \mathcal{H}_1$ solves the equation (3.92), ψ is by definition contained in the domain $D(H)$ of H and solves therefore, due to (3.90) and as $\mathcal{H}_1$ is dense in $\mathcal{H}$, the equation (3.91). Both equations, the eigenvalue equation (3.91) and its weak form (3.92), are thus completely equivalent and can be replaced by each other. Similar considerations are possible for the time-dependent Schrödinger equation (3.37).

In the forthcoming chapters we will focus our attention almost exclusively on the weak form (3.92) of the eigenvalue equation that is –one might believe it or not– mathematically much easier to handle than the original form (3.91) and fits perfectly into the framework of the L_2-theory of elliptic partial differential equations.

3.6 The Quantum Mechanics of Multi-Particle Systems

So far we have only considered single, isolated particles moving freely in space or inside an external potential as in the case of the harmonic oscillator. Let us now assume that we have a finite collection of N such particles with the spaces $L_2(\Omega_i)$ as system Hilbert spaces. The Hilbert space describing the system that is composed of these particles is then the tensor product of these Hilbert spaces or a subspace of this space, i.e., in the given case a space of square integrable wave functions

$$\psi : \Omega_1 \times \ldots \times \Omega_N \to \mathbb{C} : (\xi_1, \ldots, \xi_N) \to \psi(\xi_1, \ldots, \xi_N). \qquad (3.93)$$

From the point of view of mathematics, this is of course another postulate that can in a strict sense not be derived from anything else, but is motivated by the statistical interpretation of the wave functions and particular of the quantity $|\psi|^2$ as a probability density. Assume that the particles can be distinguished from each other. The probability to find the particles i in the subsets Ω_i' of Ω_i is then the integral of this probability density over the cartesian product $\Omega_1' \times \ldots \times \Omega_N'$ of these Ω_i'. If

$$\psi(\xi_1, \ldots, \xi_N) = \prod_{i=1}^{N} \phi_i(\xi_i), \qquad (3.94)$$

which means that the particles do not interact and are thus statistically independent of each other, this probability is the product of the individual probabilities

$$\int_{\Omega_i'} |\phi_i(\xi_i)|^2 \, d\xi_i, \qquad (3.95)$$

as the statistical interpretation requires. The space of the square integrable wave functions (3.93) is the completion of the space spanned by the products (3.94) of the square integrable functions ϕ_i from the configuration spaces Ω_i to $\mathbb{C}$.

Quantum mechanical particles of the same type, like electrons, cannot be distinguished from each other by any means or experiment. This is both a physical statement and a mathematical postulate that needs to be specified precisely. It has striking consequences for the form of the physically admissible wave functions and of the Hilbert spaces that describe such systems of indistinguishable particles.

To understand these consequences, we have to recall that an observable quantity like momentum or energy is described in quantum mechanics by a self-adjoint operator A and that the inner product $(\psi, A\psi)$ represents the expectation value for the outcome of a measurement of this quantity in the physical state described by the normalized wave function ψ. The question is whether two distinct normalized wave functions, that is, unit vectors in the Hilbert space of the system, can represent the same physical state and how such wave function are then related.

At least a necessary condition that two normalized elements or unit vectors ψ and ψ' in the system Hilbert space $\mathcal{H}$ describe the same physical state is surely that $(\psi, A\psi) = (\psi', A\psi')$ for all self-adjoint operators $A : D(A) \subseteq \mathcal{H} \to \mathcal{H}$ whose domain $D(A)$ contains both ψ and ψ', that is, that the expectation values of all possible observables coincide. This requirement fixes such states almost completely:

Lemma 3.4. *Let f and g be given unit vectors in the complex Hilbert space $\mathcal{H}$ and assume that $(f, Sf) = (g, Sg)$ for all bounded symmetric operators $S : \mathcal{H} \to \mathcal{H}$. Then there exists a real number θ such that $g = e^{i\theta} f$ and vice versa.*

Proof. The proof is an easy exercise in linear algebra. Assume that f and g are linearly independent, that is, span a two-dimensional subspace. The vectors f and

$$h = g - (g, f)f \neq 0$$

form then an orthogonal basis of this subspace and every vector

$$v = \alpha f + \beta h + v'$$

in $\mathcal{H}$ can be uniquely decomposed into a linear combination of f and h and a further vector v' that is orthogonal to these two. We consider the symmetric operator

$$Sv = \alpha f + 2\beta h$$

defined in terms of this decomposition. A short calculation shows that

$$1 = (f, Sf) = (g, Sg) = 2 - |(f, g)|^2$$

or, since f and g are unit vectors, $|(f, g)| = \|f\| \|g\|$. This is a contradiction to the linear independence of f and g. Thus $g = e^{i\theta} f$ for some real number θ. $\qquad\square$

Wave functions that describe the same physical state can therefore differ at most by a constant phase shift $\psi \to e^{i\theta} \psi$, θ a real number. Wave functions that differ by such a phase shift lead to the same expectation values of observable quantities.

In view of this discussion the requirements on the wave functions describing a system of indistinguishable particles are rather obvious and can be formulated in terms of the operations that formally exchange the single particles:

Postulate 3. The Hilbert space of a system of N indistinguishable particles with system Hilbert space $L_2(\Omega)$ consists of complex-valued, square integrable functions

$$\psi : \Omega \times \ldots \times \Omega \to \mathbb{C} : (\xi_1, \ldots, \xi_N) \to \psi(\xi_1, \ldots, \xi_N) \qquad (3.96)$$

on the N-fold cartesian product of Ω, that is, is a subspace of $L_2(\Omega^N)$. For every ψ in this space and every permutation P of the arguments ξ_i, the function $\xi \to \psi(P\xi)$ is also in this space, and moreover it differs from ψ at most by a constant phase shift.

This postulate can be rather easily translated into a symmetry condition on the wave functions that governs the quantum mechanics of multi-particle systems:

Theorem 3.7. *The Hilbert space describing a system of indistinguishable particles either consists completely of antisymmetric wave functions, functions ψ for which*

$$\psi(P\xi) = \text{sign}(P)\psi(\xi) \tag{3.97}$$

holds for all permutations P of the components $\xi_1, \ldots, \xi_N$ of ξ, that is, of the single particles, or only of symmetric wave functions, wave functions for which

$$\psi(P\xi) = \psi(\xi) \tag{3.98}$$

holds for all permutations P of the arguments.

Proof. We first fix a single wave function ψ and show that it must be symmetric or antisymmetric. Let $\alpha(P)$ be the phase shift assigned to the permutation P, that is, let

$$\psi(P\xi) = \alpha(P)\psi(\xi)$$

for all arguments $\xi \in \Omega^N$. For all permutations P and Q then necessarily

$$\alpha(PQ) = \alpha(P)\alpha(Q).$$

Next we consider transpositions, permutations that exchange two components. Transposition are conjugate to each other, which means that for every pair of transpositions T and T' there is a permutation P with $T' = P^{-1}TP$, from which by the relation above $\alpha(T') = \alpha(T)$ follows. Since transpositions are self-inverse,

$$\alpha(T)^2 = \alpha(T^2) = \alpha(I) = 1.$$

Thus there remain only two cases: either $\alpha(T) = -1$ for all transpositions T and with that $\alpha(P) = \text{sign}(P)$ for all permutations P, or $\alpha(T) = 1$ for all transpositions and $\alpha(P) = 1$ for all permutations. In the first case, the given wave function ψ is antisymmetric, and in the second one symmetric.

The Hilbert space can therefore only contain symmetric and antisymmetric functions. But a sum of a symmetric and an antisymmetric function can only be symmetric if the antisymmetric part vanishes, and antisymmetric if the symmetric part vanishes. The Hilbert space must therefore either completely consist of symmetric functions, or completely of antisymmetric functions.   □

Which of the two choices is realized depends solely on the kind of particles and cannot be decided in the present framework. Particles with antisymmetric wave functions are called fermions and particles with symmetric wave functions bosons.

We are interested in electrons. Electrons have a position in space and an internal property called spin that in many respects behaves like an angular momentum. The spin σ of an electron can attain the two values $\sigma = \pm 1/2$. The configuration space of an electron is therefore not the $\mathbb{R}^3$ but the cartesian product

$$\Omega = \mathbb{R}^3 \times \{-1/2, +1/2\}. \tag{3.99}$$

The space $L_2(\Omega)$ consists of the functions $\psi : \Omega \to \mathbb{C}$ with square integrable components $x \to \psi(x, \sigma)$, $\sigma = \pm 1/2$, and is equipped with the inner product

$$(\psi, \phi) = \sum_{\sigma = \pm 1/2} \int \psi(x, \sigma) \overline{\phi(x, \sigma)} \, dx. \tag{3.100}$$

A system of N electrons is correspondingly described by wave functions

$$\psi : (\mathbb{R}^3)^N \times \{-1/2, 1/2\}^N \to \mathbb{C} : (x, \sigma) \to \psi(x, \sigma) \tag{3.101}$$

with square integrable components $x \to \psi(x, \sigma)$, with σ now a vector consisting of N spins $\sigma_i = \pm 1/2$. These wave functions are equipped with the inner product

$$(\psi, \phi) = \sum_{\sigma} \int \psi(x, \sigma) \overline{\phi(x, \sigma)} \, dx, \tag{3.102}$$

where the sum now runs over the corresponding 2^N spin vectors σ. Electrons are fermions, as all particles with half-integer spin. That is, the wave functions change their sign under a simultaneous exchange of the positions x_i and x_j and the spins σ_i and σ_j of electrons $i \neq j$. They are, in other words, antisymmetric in the sense that

$$\psi(Px, P\sigma) = \text{sign}(P)\psi(x, \sigma) \tag{3.103}$$

holds for arbitrary simultaneous permutations $x \to Px$ and $\sigma \to P\sigma$ of the electron positions and spins. This is a general version of the Pauli principle, a principle that is of fundamental importance for the physics of atoms and molecules.

The Pauli principle has stunning consequences. It entangles the electrons with each other, without the presence of any direct interaction force. A wave function (3.101) describing such a system vanishes at points (x, σ) at which $x_i = x_j$ and $\sigma_i = \sigma_j$ for indices $i \neq j$. This means that two electrons with the same spin cannot meet at the same place, a purely quantum mechanical repulsion effect that has no counterpart in classical physics and will play a decisive role in our further reasoning.

Finally we consider again the harmonic oscillator and begin with the case of a single electron. The Hamiltonian is the same as discussed in Sect. 3.3, but it acts now on wave functions ψ with two components $x \to \psi(x, \sigma)$, one for each of the two possible values of the spin. The eigenfunctions are therefore products of the known position-dependent eigenfunctions with functions χ depending only on the spin variable. These functions χ form a two-dimensional space. That is, the eigenvalues remain the old ones but their multiplicity is doubled. The Hamiltonian for a system of N electrons that move in the potential of a harmonic oscillator but do not directly interact with each other reads in dimensionless form

$$H = \sum_{i=1}^{N} \left\{ -\frac{1}{2}\Delta_i + \frac{1}{2}|x_i|^2 \right\}, \tag{3.104}$$

where Δ_i denotes the three-dimensional Laplacian acting upon the components of the position vector x_i of the electron i. The eigenfunctions of this Hamiltonian are the antisymmetric linear combinations of the products of the one-particle eigenfunctions discussed in Sect. 3.4, the so-called Slater determinants

$$\frac{1}{\sqrt{N!}} \det \left(\phi_i(x_j, \sigma_j) \right) \tag{3.105}$$

built up from them. Such Slater determinants are only different from zero when the functions ϕ_i are linearly independent of each other and, up to a possible change of sign, do not depend on their ordering. To find the ground states of the aggregate system, that is, the eigenfunctions for the minimum eigenvalue, one therefore has to fill up these orbitals consecutively with eigenfunctions of minimum possible energy, a procedure that is denoted as aufbau principle in the physical and chemical literature. One starts with the two eigenfunctions for the eigenvalue $3/2$, one corresponding to spin $-1/2$ and the other to spin $+1/2$, proceeds with the 2×3 eigenfunctions for the eigenvalue $5/2$, and so on, until all N electrons are distributed. The minimum eigenvalue in the case of 10 electrons is, for example,

$$2 \times \frac{3}{2} + 6 \times \frac{5}{2} + 2 \times \frac{7}{2} = 25$$

Since the eigenvalue $7/2$ of the one-particle operator has multiplicity 12, there are

$$\binom{12}{2} = 66$$

possibilities to choose the two orbitals of highest energy, which means that the multiplicity of the minimum eigenvalue 25 of the ten-particle operator is 66.

In reality, the electrons interact with each other. The Hamiltonians therefore no longer split into distinct parts, each acting only on the coordinates of a single electron, and the eigenfunctions can no longer be built up from one-particle eigenfunctions. For a system consisting of N electrons, they depend on $3N$ variables and have 2^N components. The challenge is to reduce the horrifying complexity of these objects to a level that comes into the reach of numerical methods.

Chapter 4
The Electronic Schrödinger Equation

Atoms, molecules, and ions are described by the Schrödinger equation for a system of charged particles that interact by Coulomb attraction and repulsion forces. As the nuclei are much heavier than the electrons, the electrons almost instantaneously follow their motion. Therefore it is usual in quantum chemistry and related fields to separate the motion of the nuclei from that of the electrons and to start from the electronic Schrödinger equation, the equation that describes the motion of a finite set of electrons in the field of a finite number of clamped nuclei, or in other words to look for the eigenvalues and eigenfunctions of the Hamilton operator

$$H = -\frac{1}{2}\sum_{i=1}^{N}\Delta_i - \sum_{i=1}^{N}\sum_{v=1}^{K}\frac{Z_v}{|x_i - a_v|} + \frac{1}{2}\sum_{\substack{i,j=1\\i\neq j}}^{N}\frac{1}{|x_i - x_j|} \tag{4.1}$$

written down here in dimensionless form or atomic units. It acts on functions with arguments $x_1,\ldots,x_N \in \mathbb{R}^3$, which are associated with the positions of the considered electrons. The $a_1,\ldots,a_K \in \mathbb{R}^3$ are the fixed positions of the nuclei and the positive values Z_v the charges of the nuclei in multiples of the electron charge.

Like the Hamilton operator for a system of electrons moving in the potential of a harmonic oscillator, the Hamilton operator (4.1) is derived via the correspondence principle from its counterpart in classical physics, the Hamilton function or total energy of a system of point-like particles in a potential field. It is again composed of two parts, a first part representing the kinetic energy of the electrons, built up from the Laplacians Δ_i acting upon their position vectors x_i, and the potential part

$$V = -\sum_{i=1}^{N}\sum_{v=1}^{K}\frac{Z_v}{|x_i - a_v|} + \frac{1}{2}\sum_{\substack{i,j=1\\i\neq j}}^{N}\frac{1}{|x_i - x_j|} \tag{4.2}$$

describing the interaction of the electrons among each other and with the nuclei. The difficulty is not only that these potentials are singular but that the electrons are coupled to each other so that the eigenfunctions are no longer products or linear combinations of products of three-dimensional one-electron eigenfunctions.

H. Yserentant, *Regularity and Approximability of Electronic Wave Functions*,
Lecture Notes in Mathematics 2000, DOI 10.1007/978-3-642-12248-4_4,
© Springer-Verlag Berlin Heidelberg 2010

The transition from the full, time-dependent Schrödinger equation taking also into account the motion of the nuclei to the electronic Schrödinger equation is a mathematically very subtle problem that is not addressed here; we refer to [78] and the literature cited therein. The present book is concerned with the study of the analytical properties of the eigenfunctions of the operator (4.1) with the aim to find points of attack to approximate them efficiently. This chapter is devoted to the precise mathematical formulation of the electronic Schrödinger equation. Our approach is based on the weak formulation of the problem outlined in Sect. 3.5.

4.1   The Hardy Inequality and the Interaction Energy

We first neglect the spin-dependence of the wave functions that will then be taken into account in the next section. Since the eigenvalues of a self-adjoint operator are always real, the electronic Schrödinger equation splits into two separate equations of the same form for the real and the imaginary part of the wave functions. We can therefore restrict ourselves in the sequel to real-valued wave functions

$$u : (\mathbb{R}^3)^N \to \mathbb{R} : (x_1, \ldots, x_N) \to u(x_1, \ldots, x_N), \tag{4.3}$$

which, of course, need to be square integrable. Their L_2-norm given by

$$\|u\|_0^2 = \int |u(x)|^2 \, dx \tag{4.4}$$

is usually normalized to one. The integral of the function $x \to |u(x)|^2$ over a subdomain of the $\mathbb{R}^{3N}$ then represents the probability that the electrons are located in this part of the configuration space and the quantity

$$-\frac{1}{2} \sum_{i=1}^N \int u \, \Delta_i u \, dx = \frac{1}{2} \sum_{i=1}^N \int |\nabla_i u|^2 \, dx, \tag{4.5}$$

provided that it exists, the expectation value of the kinetic energy. That is, wave functions must possess first-order weak derivatives and the H^1-seminorm given by

$$|u|_1^2 = \int |(\nabla u)(x)|^2 \, dx \tag{4.6}$$

must remain finite. The solution space of the eigenvalue problem must be a subspace of the Hilbert space $H^1(\mathbb{R}^{3N})$ or briefly H^1, the space that consists of the square integrable functions (4.3) with square integrable first-order weak partial derivatives and that is equipped with the H^1-norm given by the expression

$$\|u\|_1^2 = \|u\|_0^2 + |u|_1^2. \tag{4.7}$$

In Sect. 2.1 we introduced the space $\mathscr{D}$ of the infinitely differentiable functions with compact support. From Sect. 2.3 we know that the functions in $\mathscr{D}$ form a dense subset of H^1 and H^1 can thus be considered as completion of $\mathscr{D}$ under the norm (4.7).

The rest of this section is based on a classical inequality, the Hardy inequality for functions defined on $\mathbb{R}^3$. Hardy-type inequalities play a central role in this work.

Lemma 4.1. *For all infinitely differentiable functions v in the variable $x \in \mathbb{R}^3$ that have a compact support,*

$$\int \frac{1}{|x|^2} v^2 \, dx \le 4 \int |\nabla v|^2 \, dx. \qquad (4.8)$$

Proof. Let $d(x) = |x|$ for abbreviation. To avoid any difficulty, we assume at first that v vanishes on a neighborhood of the origin. Using the relation

$$\frac{1}{d^2} = -\nabla\left(\frac{1}{d}\right) \cdot \nabla d,$$

integration by parts then yields

$$\int \frac{1}{d^2} v^2 \, dx = \int \frac{1}{d} \nabla \cdot (v^2 \nabla d) \, dx$$

or, using $\Delta d = 2/d$ and resolving for the left-hand side, the representation

$$\int \frac{1}{d^2} v^2 \, dx = -2 \int \frac{1}{d} v \nabla d \cdot \nabla v \, dx$$

of the integral to be estimated. The Cauchy-Schwarz inequality yields

$$\int \frac{1}{d^2} v^2 \, dx \le 2 \left(\int \frac{1}{d^2} v^2 \, dx \right)^{1/2} \left(\int |\nabla d \cdot \nabla v|^2 \, dx \right)^{1/2}$$

or, using $|\nabla d| = 1$, the estimate (4.8) for functions v vanishing near the origin. To complete the proof, let $\omega : \mathbb{R}^3 \to [0, 1]$ be an infinitely differentiable cut-off function with $\omega(x) = 0$ for $|x| \le 1/2$ and with $\omega(x) = 1$ for $|x| \ge 1$. Set

$$v_k(x) = \omega(kx)v(x).$$

The estimate (4.8) then holds for the functions v_k as just proved. Using

$$|\omega(kx)| \le 1, \quad |k(\nabla \omega)(kx)| \le \frac{c}{|x|}$$

with a constant c independent of k and the local integrability of

$$x \to \frac{1}{|x|^2},$$

the proposition follows with help of the dominated convergence theorem. □

The Hardy inequality (4.8) first serves to estimate terms involving the potential

$$V(x) = -\sum_{i=1}^{N}\sum_{v=1}^{K}\frac{Z_v}{|x_i - a_v|} + \frac{1}{2}\sum_{\substack{i,j=1 \\ i \neq j}}^{N}\frac{1}{|x_i - x_j|} \qquad (4.9)$$

in the Hamilton operator (4.1) that is composed of the nucleus-electron interaction potential, the first term in (4.9), and the electron-electron interaction potential. Let Z denote the total charge of the nuclei, the sum of the charges Z_v, and set

$$\theta(N,Z) = \sqrt{N}\,\max(N,Z). \qquad (4.10)$$

A simple calculation on the basis of the Hardy inequality (4.8), Fubini's theorem, and the Cauchy-Schwarz inequality then yields our first important estimate:

Theorem 4.1. *The functions u and v in $\mathscr{D}$ satisfy the estimate*

$$\int Vuv\,dx \leq 3\theta(N,Z)\|u\|_0\|v\|_1. \qquad (4.11)$$

Next we write the Hamilton operator (4.1) in the form

$$H = -\frac{1}{2}\Delta + V \qquad (4.12)$$

and introduce the bilinear form

$$a(u,v) = (Hu,v) \qquad (4.13)$$

on $\mathscr{D}$, where $(\,,\,)$ denotes the L_2-inner product. Since

$$(-\Delta u, v) = \int \nabla u \cdot \nabla v\,dx, \qquad (4.14)$$

there exists, by Theorem 4.1, a constant M depending on N and on Z with

$$a(u,v) \leq M\|u\|_1\|v\|_1 \qquad (4.15)$$

for all $u, v \in \mathscr{D}$. The bilinear form (4.13) can therefore be extended to a bounded, symmetric bilinear form on H^1. Furthermore, for $\mu \geq 9\theta^2 + 1/4$ and all $u, v \in H^1$,

$$a(u, u) + \mu(u, u) \geq \frac{1}{4} \|u\|_1^2. \tag{4.16}$$

Neglecting the spin, the Sobolev space H^1 would therefore be the proper Hilbert space associated with the given system of electrons and the value $a(u, u)$ the expectation value of the total energy in the state described by the normed wave function $u \in H^1$. A function $u \neq 0$ in H^1 is an eigenfunction of the Hamilton operator (4.1) or (4.12), and the real number λ the associated eigenvalue, if the relation

$$a(u, \chi) = \lambda(u, \chi) \tag{4.17}$$

holds for all $\chi \in H^1$. That is, we consider weak solutions of the eigenvalue equation

$$Hu = \lambda u, \tag{4.18}$$

in the same way as this has been discussed in Sect. 3.5 in conjunction with the Friedrichs extension and as one defines weak solutions of boundary value problems. The relation (4.16) shows that the eigenvalues λ are bounded from below.

4.2   Spin and the Pauli Principle

As described in Sect. 3.6, electrons have an internal property called spin that behaves similar to angular momentum. Although spin does not explicitly appear in the electronic Schrödinger equation, it influences the structure of atoms and molecules decisively. The purpose of this section is to explain how spin can be incorporated into the variational framework. The spin of an electron can attain the two half-integer values $\pm 1/2$. Correspondingly, the true wave functions are of the form

$$\psi : (\mathbb{R}^3)^N \times \{-1/2, 1/2\}^N \to \mathbb{R} : (x, \sigma) \to \psi(x, \sigma), \tag{4.19}$$

that is, depend not only on the positions $x_i \in \mathbb{R}^3$, but also on the spins $\sigma_i = \pm 1/2$ of the electrons. The Pauli principle, one of the fundamental principles of quantum mechanics, states that only those eigenfunctions are admissible that change their sign under a simultaneous exchange of the positions x_i and x_j and the spins σ_i and σ_j of two electrons i and j, that is, are antisymmetric in the sense that

$$\psi(Px, P\sigma) = \text{sign}(P)\psi(x, \sigma) \tag{4.20}$$

holds for arbitrary simultaneous permutations $x \to Px$ and $\sigma \to P\sigma$ of the electron positions and spins. The Pauli principle forces the admissible wave functions to vanish where $x_i = x_j$ and $\sigma_i = \sigma_j$ for $i \neq j$, that is, that the probability that two electrons i and j with the same spin meet is zero. The admissible solutions of the scalar Schrödinger equation (4.17) are those that are components

$$u : (\mathbb{R}^3)^N \to \mathbb{R} : x \to \psi(x, \sigma) \tag{4.21}$$

of an antisymmetric wave function (4.19). To clarify these relations and deduce (4.17) from the full equation incorporating spin, we introduce the bilinear forms

$$B(\psi, \psi') = \sum_\sigma a\big(\psi(\cdot, \sigma), \psi'(\cdot, \sigma)\big), \tag{4.22}$$

$$(\psi, \psi') = \sum_\sigma \big(\psi(\cdot, \sigma), \psi'(\cdot, \sigma)\big) \tag{4.23}$$

on the spaces of functions (4.19) with components in H^1, respectively, L_2 where the sums extend over the 2^N possible spin vectors σ. The quantity $B(\psi, \psi)$ represents the expectation value of the total energy for normed ψ and B is thus the bilinear form that is induced by the complete Hamilton operator of the system, the operator whose eigenvalues and eigenfunctions are sought. An antisymmetric function ψ with components in H^1 is a solution of the full problem if and only if

$$B(\psi, \psi') = \lambda(\psi, \psi') \tag{4.24}$$

for all test functions ψ' of this kind. This eigenvalue problem decouples into eigenvalue problems for the components of the eigenfunctions ψ due to the fact that the bilinear form (4.13) is invariant under permutations of the positions x_i, i.e., that

$$a\big(u(P\cdot), v(P\cdot)\big) = a(u, v) \tag{4.25}$$

holds for all such permutations P and all functions $u, v \in H^1$. This property translates into a statement on the antisymmetrization operator $\mathscr{A}$ given by

$$(\mathscr{A}\psi)(x, \sigma) = \frac{1}{N!} \sum_P \operatorname{sign}(P)\psi(Px, P\sigma) \tag{4.26}$$

where the sum extends over the $N!$ possible permutations of the electrons. It maps an arbitrary function (4.19) into an antisymmetric function and reproduces antisymmetric functions. For all functions (4.19) with components in H^1 respectively L_2,

$$B(\psi, \mathscr{A}\psi') = B(\mathscr{A}\psi, \psi'), \quad (\psi, \mathscr{A}\psi') = (\mathscr{A}\psi, \psi'). \tag{4.27}$$

Theorem 4.2. *An antisymmetric function ψ with components in H^1 satisfies the eigenvalue equation (4.24) if and only if its components solve the equations*

$$a\big(\psi(\cdot, \sigma), v\big) = \lambda\big(\psi(\cdot, \sigma), v\big), \quad v \in H^1. \tag{4.28}$$

Proof. Let $\delta(\eta,\sigma) = 1$ if $\eta = \sigma$ and $\delta(\eta,\sigma) = 0$ otherwise. Every function (4.19) with components in H^1 can then be written as

$$\psi(x,\eta) = \sum_\sigma \psi(x,\sigma)\delta(\eta,\sigma),$$

that is, as a linear combination of functions of the form

$$\psi'(x,\eta) = v(x)\delta(\eta,\sigma)$$

with $v \in H^1$ and some given σ, and every antisymmetric function therefore as a linear combination of antisymmetrized functions of this form. It suffices therefore to restrict oneself to test functions $\mathscr{A}\psi'$ where ψ' is a function of the given form. Let ψ now be an arbitrary antisymmetric function with components in H^1. Then

$$B(\psi,\mathscr{A}\psi') = B(\mathscr{A}\psi,\psi') = B(\psi,\psi') = a(\psi(\cdot,\sigma),v),$$
$$(\psi,\mathscr{A}\psi') = (\mathscr{A}\psi,\psi') = (\psi,\psi') = (\psi(\cdot,\sigma),v),$$

from which the proposition follows. □

The components of the solutions ψ of the full equation (4.24) are therefore indeed solutions of the scalar equation (4.17). To characterize these components, let $\mathscr{D}(\sigma)$ denote the space of all functions $u \in \mathscr{D}$ with

$$u(Px) = \text{sign}(P)u(x) \qquad (4.29)$$

for all permutations P that leave σ invariant and let $L_2(\sigma)$ and $H^1(\sigma)$ be the closure of $\mathscr{D}(\sigma)$ in the corresponding spaces.

Theorem 4.3. *A function in $\mathscr{D}$ is the component (4.21) of an antisymmetric function (4.19) with components in $\mathscr{D}$ if and only if it belongs to $\mathscr{D}(\sigma)$. The corresponding statement holds for functions with components in L_2 and H^1, respectively.*

Proof. If ψ is antisymmetric, $u(x) = \psi(x,\sigma)$, and $P\sigma = \sigma$, then

$$u(Px) = \psi(Px,\sigma) = \psi(Px,P\sigma) = \text{sign}(P)\psi(x,\sigma) = \text{sign}(P)u(x),$$

so that the components (4.21) of an antisymmetric function are of the form (4.29). A function u satisfying (4.29) is conversely the component $u(x) = \psi(x,\sigma)$ of

$$\psi(x,\eta) = \frac{\sum_P \text{sign}(P)u(Px)\delta(P\eta,\sigma)}{\sum_P \delta(P\sigma,\sigma)},$$

and can thus be recovered from an antisymmetric function. □

The components $u = \psi(\cdot, \sigma)$ in $H^1(\sigma)$ of the full, spin-dependent eigenfunctions ψ solve, by Theorem 4.2, particularly the reduced eigenvalue equation

$$a(u, v) = \lambda(u, v), \quad v \in H^1(\sigma), \tag{4.30}$$

that results from (4.28) replacing the test space H^1 by its subspace $H^1(\sigma)$. From the solutions of these equations, one can conversely recover solutions of the full equation (4.24) combining all 2^N components of the eigenfunctions ψ.

Theorem 4.4. *If the function $u \neq 0$ in $H^1(\sigma)$ solves the eigenvalue equation (4.30) reduced to the space $H^1(\sigma)$, the antisymmetric function $\psi \neq 0$ defined by*

$$\psi(x, \eta) = \frac{1}{N!} \sum_P \text{sign}(P) u(Px) \delta(P\eta, \sigma) \tag{4.31}$$

solves the full equation (4.24) and the function u itself solves the original equation

$$a(u, v) = \lambda(u, v), \quad v \in H^1. \tag{4.32}$$

Proof. Let ψ' be an antisymmetric function with components in H^1. Its component $x \to \psi'(x, \sigma)$ then belongs to $H^1(\sigma)$. Since, as in the proof of Theorem 4.2,

$$B(\psi, \psi') = B(\psi', \psi) = a(\psi'(\cdot, \sigma), u) = a(u, \psi'(\cdot, \sigma)),$$
$$(\psi, \psi') = (\psi', \psi) = (\psi'(\cdot, \sigma), u) = (u, \psi'(\cdot, \sigma)),$$

the function (4.31) therefore solves the equation (4.24) for the complete, spin-dependent wave functions. As $u(x) = \text{sign}(P)u(Px)$ whenever P fixes σ, u is a constant multiple of the function $\psi(\cdot, \sigma)$. The rest follows from Theorem 4.2. □

With that the circle is closed. Since the functions $u \in H^1(\sigma)$ and $\tilde{u}(x) = u(Q^{-1}x)$ in $H^1(Q\sigma)$ generate, up to a possible change of sign, the same function (4.31) for arbitrary permutations Q of the electrons, and since $\tilde{u} \in H^1(Q\sigma)$ solves the equation

$$a(\tilde{u}, \tilde{v}) = \lambda(\tilde{u}, \tilde{v}), \quad \tilde{v} \in H^1(Q\sigma), \tag{4.33}$$

if and only if u solves (4.30), one can restrict oneself to the reduced equations (4.30) on the $\lfloor N/2 \rfloor$ essentially different spaces $H^1(\sigma)$ instead of solving the system (4.24) for the 2^N components of a wave function (4.19) directly. Every solution of such a reduced equation also solves the eigenvalue problem (4.17) on the bigger space H^1.

Chapter 5
Spectrum and Exponential Decay

In this chapter we begin to study the solutions of the electronic Schrödinger equation and compile and prove some basic, for the most part well-known, facts about its solutions in suitable form. Parts of this chapter are strongly influenced by Agmon's monograph [3] on the exponential decay of the solutions of second-order elliptic equations. Starting point are two constants associated with the solution spaces introduced in the previous chapter, the minimum energy that the given system can attain and the ionization threshold. Both constants are intimately connected with the spectral properties of the Hamilton operator and are introduced in the first section of this chapter. The second section deals with some notions and simple results from spectral theory that are rewritten here in terms of bilinear forms as they underly the weak form of the Schrödinger equation. The weak form of the equation will not only be the starting point of the regularity theory that we will develop later, but is also the basis for many approximation methods of variational type, from the basic Ritz method discussed in the third section to the many variants and extensions of the Hartree-Fock method. Our exposition is based on simple, elementary properties of Hilbert spaces like the projection theorem, the Riesz representation theorem, or the fact that every bounded sequence contains a weakly convergent subsequence. Hence only a minimum of prerequisites from functional analysis is required. For a comprehensive treatment of spectral theory and its application to quantum mechanics, we refer to texts like [44, 69–71], or [87, 88]. We finally show, in the fourth section, that the essential spectrum of the electronic Schrödinger operator is non-empty and that the ionization threshold represents its lower bound. We will assume that the minimum energy is located below the ionization threshold. It is then an eigenvalue, the ground state energy. The corresponding eigenfunctions are the ground states. The knowledge of the ground states and particularly of the ground state energy is of main interest in quantum chemistry. The last section is devoted to the exponential decay of the eigenfunctions for eigenvalues below the ionization threshold, a result that goes back to O'Connor [20] and has later been substantially refined [3]. In contrast to many other presentations the symmetry properties of the wave functions enforced by the Pauli principle are hereby carefully taken into account.

H. Yserentant, *Regularity and Approximability of Electronic Wave Functions*,
Lecture Notes in Mathematics 2000, DOI 10.1007/978-3-642-12248-4_5,
© Springer-Verlag Berlin Heidelberg 2010

5.1 The Minimum Energy and the Ionization Threshold

Recall that we denoted by $\mathscr{D}$ the space of the infinitely differentiable functions with bounded support and that the space $\mathscr{D}(\sigma)$ consists of the functions in $\mathscr{D}$ that are antisymmetric under the permutations of the positions of the electrons that leave the given spin vector σ invariant. The Sobolev space H^1 is the completion of $\mathscr{D}$ under the norm (4.7) and the space $H^1(\sigma)$ the closure of $\mathscr{D}(\sigma)$ in H^1. Let $a(u,v)$ be the extension of the bilinear form (4.13) from $\mathscr{D}$ to H^1. From (4.16) we know that the total energy is bounded from below. Hence we are allowed to define the constant

$$\Lambda(\sigma) = \inf\left\{a(u,u)\,\big|\,u \in \mathscr{D}(\sigma), \|u\|_0 = 1\right\}, \tag{5.1}$$

the minimum energy that the system can attain with the given distribution of spins. Its counterpart is the ionization threshold. To prepare its definition let

$$\Sigma(R,\sigma) = \inf\left\{a(u,u)\,\big|\,u \in \mathscr{D}(\sigma), \|u\|_0 = 1, u(x) = 0 \text{ for } |x| \leq R\right\}. \tag{5.2}$$

Lemma 5.1. *The constants $\Sigma(R,\sigma)$ are bounded from above by the value zero.*

Proof. Let $u \neq 0$ in $\mathscr{D}(\sigma)$ be a normed infinitely differentiable function that vanishes on the ball of radius 1 around the origin of the $\mathbb{R}^{3N}$. The rescaled functions

$$u_R(x) = \frac{1}{R^{3N/2}}\,u\left(\frac{x}{R}\right)$$

then have L_2-norm 1, too, and vanish on the ball of radius R around the origin. Therefore, by the definition (5.1) of the constant $\Sigma(R,\sigma)$,

$$\Sigma(R,\sigma) \leq a(u_R, u_R).$$

At this place, the particular properties of the given potential enter. By Theorem 4.1,

$$a(u_R, u_R) \leq \frac{1}{2}\,|u_R|_1^2 + 3\theta(N,Z)\,\|u_R\|_0 |u_R|_1.$$

This estimate can be rewritten in terms of the original function u using the relations

$$\|u_R\|_0 = \|u\|_0, \quad |u_R|_1 = \frac{1}{R}|u|_1.$$

For arbitrarily given $\varepsilon > 0$ and R chosen sufficiently large, therefore $\Sigma(R,\sigma) \leq \varepsilon$. As the $\Sigma(R,\sigma)$ are monotonely increasing in R, $\Sigma(R,\sigma) \leq \varepsilon$ for all $R > 0$ follows. Since ε can be chosen arbitrarily small, this implies the proposition. $\qquad\square$

As the $\Sigma(R,\sigma)$ are monotonely increasing in R, we can therefore define the constant

$$\Sigma(\sigma) = \lim_{R\to\infty} \Sigma(R,\sigma) \le 0, \tag{5.3}$$

the energy threshold above which at least one electron has moved arbitrarily far away from the nuclei, the ionization threshold. As one knows from [65], and as we will show in the fourth section, it is closely linked to the spectral properties of the Hamilton operator (4.1), respectively the corresponding bilinear form $a(u,v)$, and represents the infimum of the essential spectrum. Our main assumption is that

$$\Lambda(\sigma) < \Sigma(\sigma), \tag{5.4}$$

that is, that it is energetically more advantageous for the electrons to stay in the vicinity of the nuclei than to fade away at infinity. As we will see later, this assumption implies that the minimum energy (5.1) is an isolated eigenvalue and that the corresponding eigenfunctions, the ground states of the system, decay exponentially. The condition thus means that the nuclei can bind all electrons, which evidently does not always need to be the case, but of course holds for stable atoms and molecules.

5.2 Discrete and Essential Spectrum

The purpose of this section is to introduce some basic concepts and facts from spectral theory that are here rewritten in terms of bilinear forms as they are considered in the L_2-theory of linear elliptic differential equations. We start from an abstract framework with two real Hilbert spaces $\mathcal{H}_0$ and $\mathcal{H}_1 \subseteq \mathcal{H}_0$. Let $(\cdot,\cdot)$ denote the inner product and $\|\cdot\|_0$ the induced norm on $\mathcal{H}_0$ and $\|\cdot\|_1$ the norm on $\mathcal{H}_1$. We suppose that $\mathcal{H}_1$ is a dense subspace of $\mathcal{H}_0$ and that there exists a constant c with

$$\|u\|_0 \le c\,\|u\|_1, \quad u \in \mathcal{H}_1, \tag{5.5}$$

that is, $\mathcal{H}_1$ is densely embedded in $\mathcal{H}_0$. Furthermore, let

$$a: \mathcal{H}_1 \times \mathcal{H}_1 \to \mathbb{R}: u,v \to a(u,v) \tag{5.6}$$

be a symmetric bilinear form that is bounded in the sense that

$$a(u,v) \le M\,\|u\|_1\|v\|_1, \quad u,v \in \mathcal{H}_1, \tag{5.7}$$

and coercive in the sense that there is a constant $\delta > 0$ with

$$a(u,u) \ge \delta\,\|u\|_1^2, \quad u \in \mathcal{H}_1. \tag{5.8}$$

These properties imply that $a(u,v)$ is an inner product on $\mathcal{H}_1$ that induces a norm which is equivalent to the original norm and can substitute it.

In the case that we have in mind $\mathcal{H}_0$ is the Hilbert space L_2, respectively one of its subspaces $L_2(\sigma)$ with the corresponding symmetries built in, and $\mathcal{H}_1$ the Hilbert space H^1 of the square integrable, one times weakly differentiable functions from $\mathbb{R}^{3N}$ to $\mathbb{R}$, respectively its corresponding subspace $H^1(\sigma)$. The condition (5.8) is formally more restrictive than the condition (4.16) that the bilinear form (4.13) satisfies. It is, however, possible to replace bilinear forms like (4.13) by shifted versions as in (4.16) that satisfy (5.8) since this results only in a shift of the spectrum.

Eigenvalues and eigenvectors (or eigenfunctions in concrete applications) are defined in weak sense, in the same way as weak solutions of differential equations.

Definition 5.1. An eigenvalue λ of the bilinear form (5.6) is a real number for which there exists an element $u \in \mathcal{H}_1$ that is different from zero and for which

$$a(u,v) = \lambda(u,v), \quad v \in \mathcal{H}_1. \tag{5.9}$$

Every such u is called an eigenvector for the eigenvalue λ. The linear subspace $\mathcal{E}_\lambda$ consisting of these eigenvectors is the corresponding eigenspace. The multiplicity of the eigenvalue λ is the dimension of this eigenspace.

The problem is that, unlike the finite dimensional case, the fact that the number λ is not an eigenvalue does not necessarily mean that the equation

$$a(u,v) - \lambda(u,v) = (f,v), \quad v \in \mathcal{H}_1, \tag{5.10}$$

possesses a unique solution $u \in \mathcal{H}_1$ depending continuously on the data $f \in \mathcal{H}_0$.

Definition 5.2. A real number λ belongs to the resolvent of the bilinear form (5.6) if and only if the equation (5.10) possesses a unique solution $u \in \mathcal{H}_1$ for all given $f \in \mathcal{H}_0$ that depends continuously on the data, that is, if the linear mapping

$$R_\lambda : \mathcal{H}_0 \to \mathcal{H}_1 : f \to u =: R_\lambda f \tag{5.11}$$

is bounded. The values λ which do not belong to the resolvent form its spectrum.

The spectrum obviously contains the eigenvalues but can be much larger, which is the case with the bilinear forms induced by the Hamilton operators of atoms and molecules. It should be noted that, because of the identity

$$a(R_\lambda f, R_\lambda f) = \lambda(R_\lambda f, R_\lambda f) + (f, R_\lambda f) \tag{5.12}$$

and the coercivity (5.8) of the bilinear form, it suffices to require that the resolvent mapping (5.11) is bounded as a mapping from $\mathcal{H}_0$ to $\mathcal{H}_0$. Because of

$$(f, R_\lambda g) = a(R_\lambda f, R_\lambda g) - \lambda(R_\lambda f, R_\lambda g) \tag{5.13}$$

the resolvent mappings are symmetric in the sense that

$$(R_\lambda f, g) = (f, R_\lambda g), \quad f, g \in \mathcal{H}_0. \tag{5.14}$$

The spectrum of the bilinear form (5.6) is bounded from below. A first lower bound can be given in terms of the constants from (5.5) and (5.8).

Theorem 5.1. *All real numbers $\lambda < \delta/c^2$ belong to the resolvent of the bilinear form; its spectrum is therefore a subset of the interval $\lambda \geq \delta/c^2 > 0$.*

Proof. For $\lambda \leq 0$, the coercivity (5.8) implies

$$a(u,u) - \lambda(u,u) \geq a(u,u) \geq \delta\|u\|_1^2,$$

and for $\lambda > 0$ correspondingly

$$a(u,u) - \lambda(u,u) \geq (\delta - \lambda c^2)\|u\|_1^2.$$

The shifted bilinear form

$$u,v \;\to\; a(u,u) - \lambda(u,u)$$

is therefore coercive for $\lambda < \delta/c^2$. The proposition thus follows from the Riesz representation theorem applied to this bilinear form as the inner product on $\mathcal{H}_1$. $\square$

In particular there is a symmetric bounded linear operator $G : \mathcal{H}_0 \to \mathcal{H}_1$ with

$$a(Gf,v) = (f,v), \quad v \in \mathcal{H}_1, \tag{5.15}$$

the resolvent mapping (5.11) for $\lambda = 0$. A given element $u \in \mathcal{H}_1$ is an eigenvector of the bilinear form (5.6) for the eigenvalue λ if and only if

$$u - \lambda Gu = 0. \tag{5.16}$$

That is, in view of Theorem 5.1, the eigenvalues and eigenvectors of the bilinear form and of the linear mapping G correspond to each other.

Since G is injective, G has an inverse A with the range $D(A)$ of G as domain. For all $u \in D(A)$, $Au \in \mathcal{H}_0$ is characterized by the relation $(Au,v) = a(u,v)$ for $v \in \mathcal{H}_1$. It can be shown that A is self-adjoint and that the spectrum of A and of the bilinear form coincide. The operator A and the bilinear form determine each other. In the case in that we are mainly interested, A is the self-adjoint extension of the given Hamilton operator discussed in Sect. 3.5. We will not utilize these facts here.

Theorem 5.2. *The real number λ belongs to the resolvent of the bilinear form if and only if the bounded linear mapping*

$$I - \lambda G : \mathcal{H}_0 \to \mathcal{H}_0 \tag{5.17}$$

possesses a bounded inverse $T_\lambda : \mathcal{H}_0 \to \mathcal{H}_0$.

Proof. Let λ belong to the resolvent. We first observe that, for all given $f \in \mathcal{H}_0$,

$$a(Gf, v) - \lambda(Gf, v) = (f - \lambda Gf, v), \quad v \in \mathcal{H}_1.$$

By definition of R_λ this implies $Gf = R_\lambda(f - \lambda Gf)$ or $G = R_\lambda(I - \lambda G)$. Moreover,

$$a((I - \lambda G)R_\lambda f, v) = (f, v), \quad v \in \mathcal{H}_1,$$

from which $(I - \lambda G)R_\lambda = G$ follows. Therefore

$$I = (I + \lambda R_\lambda)(I - \lambda G) = (I - \lambda G)(I + \lambda R_\lambda),$$

that is, the operator $T_\lambda = I + \lambda R_\lambda$ is a bounded inverse of $I - \lambda G$.

Let the operator $I - \lambda G$ conversely have a bounded inverse T_λ. For $f \in \mathcal{H}_0$ given the equation (5.10) can then have at most one solution because

$$a(u, v) - \lambda(u, v) = 0, \quad v \in \mathcal{H}_1,$$

implies $u - \lambda Gu = 0$ and with that $u = 0$. On the other hand, $u = GT_\lambda f$ solves the equation (5.10). Thus λ belongs to the resolvent and $R_\lambda = GT_\lambda$. $\qquad \square$

Theorem 5.3. *The resolvent is an open and the spectrum a closed set.*

Proof. Let λ_0 belong to the resolvent and T_0 be the inverse of $I - \lambda_0 G$. We start observing that $u \in \mathcal{H}_1$ solves the equation (5.10) if and only if $u - \lambda Gu = Gf$ or

$$u = T_0 Gf + (\lambda - \lambda_0) T_0 Gu.$$

The Banach fixed point theorem guarantees that this equation possesses a unique solution depending continuously on f for all λ in a sufficiently small neighborhood of the given λ_0. All λ in this neighborhood of λ_0 belong therefore to the resolvent. Hence the resolvent is open and the spectrum correspondingly closed. $\qquad \square$

Theorem 5.4. *The value λ belongs to the spectrum of the bilinear form if and only if there exists a sequence of elements $f_n \in \mathcal{H}_0$ with*

$$\lim_{n \to \infty} \|(I - \lambda G)f_n\|_0 = 0, \quad \|f_n\|_0 = 1, \tag{5.18}$$

that is, if λ is a so-called approximate eigenvalue.

Proof. Let λ first belong to the resolvent and let T_λ be the bounded inverse of the operator $I - \lambda G$. If the vectors $r_n = (I - \lambda G)f_n$ tend then to zero in $\mathcal{H}_0$ as n goes to infinity, the same holds for the vectors $f_n = T_\lambda r_n$. Thus λ cannot be an approximate eigenvalue and the approximate eigenvalues form a part of the spectrum.

Let λ conversely belong to the spectrum. If λ is an eigenvalue, nothing has to be shown. If λ is not an eigenvalue, $I - \lambda G$ is injective. Furthermore, the range of $I - \lambda G$ is a dense subset of $\mathcal{H}_0$: Let $(f, (I - \lambda G)g) = 0$ for all $g \in \mathcal{H}_0$. Since G is symmetric, then also $((I - \lambda G)f, g) = 0$ for all $g \in \mathcal{H}_0$. This is only possible for $(I - \lambda G)f = 0$, that is, for $f = 0$ by the injectivity of $I - \lambda G$.

Therefore the inverse operator of $I - \lambda G$ mapping the range of $I - \lambda G$ back to its domain $\mathcal{H}_0$ cannot be bounded. Otherwise it could namely be extended to a bounded inverse T_λ of $I - \lambda G$ and λ would belong to the resolvent by Theorem 5.2. This means that there is a sequence of elements g_n in the range of $I - \lambda G$ such that

$$\lim_{n \to \infty} \|g_n\|_0 = 0, \quad \|(I - \lambda G)^{-1} g_n\|_0 = 1.$$

The vectors $f_n = (I - \lambda G)^{-1} g_n$ have then the properties (5.18) so that λ is indeed an approximate eigenvalue. □

Definition 5.3. An eigenvalue λ of the bilinear form (5.6) is called isolated, if there exists a constant $\vartheta > 0$ with

$$\|f\|_0 \leq \vartheta^{-1} \|(I - \lambda G)f\|_0, \quad f \in \mathscr{E}_\lambda^\perp, \tag{5.19}$$

where $\mathscr{E}_\lambda^\perp$ is the $\mathcal{H}_0$-orthogonal complement of the corresponding eigenspace

$$\mathscr{E}_\lambda = \{f \in \mathcal{H}_0 \,|\, (I - \lambda G)f = 0\} \tag{5.20}$$

of the bilinear form. The isolated eigenvalues of finite multiplicity form the discrete spectrum, the other values in the spectrum the essential spectrum.

The discrete spectrum is of special importance in the study of atoms and molecules. As we will see, it fixes the energies of the bound states and with that the frequencies of the light that the atom or molecule emits and absorbs, its spectrum.

Theorem 5.5. *All $\lambda \neq \lambda_0$ sufficiently close to an isolated eigenvalue λ_0 of finite multiplicity belong to the resolvent and all accumulation points of the spectrum to the essential spectrum.*

Proof. As G maps the corresponding eigenspace $\mathscr{E}_0$ and its orthogonal complement $\mathscr{E}_0^\perp$ into itself, the problem to solve equation (5.10) or equivalently $u - \lambda Gu = Gf$ for $f \in \mathcal{H}_0$ given splits into the subproblem to find an element $v \in \mathscr{E}_0^\perp$ with

$$v - \lambda Gv = Gf',$$

where f' denotes the orthogonal projection of f onto the subspace $\mathscr{E}_0^\perp$ of $\mathcal{H}_0$, and a corresponding subproblem on $\mathscr{E}_0$. The restriction of $I - \lambda_0 G$ to $\mathscr{E}_0^\perp$ possesses a bounded inverse $T_0' : \mathscr{E}_0^\perp \to \mathscr{E}_0^\perp$, as can be shown using an argument as in the proof of Theorem 5.4 and utilizing (5.19). The equation on $\mathscr{E}_0^\perp$ is therefore equivalent to

$$v = T_0' f' + (\lambda - \lambda_0) T_0' Gv.$$

By the Banach fixed point theorem it possesses again a unique solution for all λ sufficiently close to λ_0. There remains the subproblem on $\mathscr{E}_0$. Because

$$w - \lambda Gw = (1 - \lambda/\lambda_0) w$$

for $w \in \mathcal{E}_0$, this subproblem is solvable for $\lambda \neq \lambda_0$. The discrete spectrum contains therefore only isolated points, as the notion 'isolated eigenvalue' suggests. Since the spectrum is a closed subset of $\mathbb{R}$, this proves also the second proposition.   □

Theorem 5.6. *The value λ belongs to the essential spectrum of the bilinear form if and only if there exists a sequence of elements $f_n \in \mathcal{H}_0$ with*

$$\lim_{n \to \infty} \|(I - \lambda G) f_n\|_0 = 0, \quad \|f_n\|_0 = 1, \tag{5.21}$$

and additionally

$$f_n \to 0 \quad weakly \ in \ \mathcal{H}_0. \tag{5.22}$$

Proof. Let λ belong to the essential spectrum. If the subspace

$$\mathcal{E}_\lambda = \{ f \in \mathcal{H}_0 \,|\, (I - \lambda G) f = 0 \}$$

of $\mathcal{H}_0$ is infinite dimensional, $\mathcal{E}_\lambda$ contains a sequence of pairwise orthogonal elements f_n of norm 1. Because

$$\sum_{i=1}^{\infty} |(f_n, v)|^2 \leq \|v\|_0^2$$

for all $v \in \mathcal{H}_0$, then necessarily

$$\lim_{n \to \infty} (f_n, v) = 0$$

for all $v \in \mathcal{H}_0$. Hence the f_n converge weakly to zero and nothing is left to be done. Otherwise we decompose $\mathcal{H}_0$ into the direct sum

$$\mathcal{H}_0 = \mathcal{E}_\lambda \oplus \mathcal{E}_\lambda^\perp$$

of $\mathcal{E}_\lambda$ and its orthogonal complement $\mathcal{E}_\lambda^\perp$ in $\mathcal{H}_0$. The restriction of $I - \lambda G$ to $\mathcal{E}_\lambda^\perp$ is injective by definition. However, its inverse mapping the image of $\mathcal{E}_\lambda^\perp$ under $I - \lambda G$ back to $\mathcal{E}_\lambda^\perp$ cannot be bounded because λ would then not belong to the essential spectrum. Therefore there exists a sequence of elements $f_n \in \mathcal{E}_\lambda^\perp$ with

$$\lim_{n \to \infty} \|(I - \lambda G) f_n\|_0 = 0, \quad \|f_n\|_0 = 1.$$

As every bounded sequence in a Hilbert space contains a weakly convergent subsequence, we can assume that the f_n converge weakly in $\mathcal{H}_0$ to a limit element f. As

$$(f, v) = \lim_{n \to \infty} (f_n, v) = 0, \quad v \in \mathcal{E}_\lambda,$$

this f belongs itself to the orthogonal complement $\mathcal{E}_\lambda^\perp$ of $\mathcal{E}_\lambda$. Since

$$((I - \lambda G) f, g) = \lim_{n \to \infty} ((I - \lambda G) f_n, g) = 0$$

for all $g \in \mathscr{H}_0$, f is also contained in $\mathscr{E}_\lambda$. Therefore $f = 0$, and we have found a sequence of elements in $\mathscr{H}_0$ that satisfy both (5.21) and (5.22).

Conversely, let λ be an isolated eigenvalue of finite multiplicity and let (f_n) be a sequence of elements satisfying (5.21) and (5.22). Decompose the f_n as

$$f_n = v_n + w_n, \quad v_n \in \mathscr{E}_\lambda, w_n \in \mathscr{E}_\lambda^\perp.$$

Since $(I - \lambda G)v_n = 0$, by condition (5.19)

$$\|w_n\|_0 \leq \vartheta^{-1}\|(I - \lambda G)w_n\|_0 = \vartheta^{-1}\|(I - \lambda G)f_n\|_0$$

so that $w_n \to 0$ strongly. As $f_n \to 0$ weakly by assumption, this means $v_n \to 0$ weakly. Because $\mathscr{E}_\lambda$ is finite dimensional, this implies $v_n \to 0$ strongly. But then also $f_n \to 0$ strongly, which contradicts $\|f_n\|_0 = 1$. □

There is a simple, but very useful corollary from Theorem 5.6 that often plays an important role in dealing with the essential spectrum.

Corollary 5.1. *For every λ in the essential spectrum there exist $u_n \in \mathscr{H}_1$ with*

$$\|u_n\|_0 = 1, \quad u_n \to 0 \quad \text{weakly in } \mathscr{H}_0, \tag{5.23}$$

$$\lim_{n\to\infty} a(u_n, u_n) = \lambda. \tag{5.24}$$

Proof. Choosing the u_n proportional to Gf_n with the f_n from Theorem 5.6,

$$a(u_n, u_n) = \lambda + \frac{(f_n - \lambda Gf_n, Gf_n)}{\|Gf_n\|_0^2} \to \lambda. \qquad \square$$

By calculating the directional derivatives one can easily recognize that the eigenvectors are the stationary points of the Rayleigh quotient

$$u \to \frac{a(u, u)}{(u, u)}, \tag{5.25}$$

and that at an eigenvector u the Rayleigh quotient attains the corresponding eigenvalue λ. In particular the minimum of the Rayleigh quotient is the minimum eigenvalue in finite space dimensions. The goal is to transfer these properties to the infinite dimensional case. The situation is much more subtle there because it is not even a priori clear whether the Rayleigh quotient attains its minimum. The most general result, at the same time demonstrating that the spectrum is never empty, is:

Theorem 5.7. *The constant*

$$\Lambda = \inf\{a(u, u) \mid u \in \mathscr{H}_1, \|u\|_0 = 1\} \tag{5.26}$$

belongs to the spectrum and represents its infimum.

Proof. The range of G is a dense subspace of $\mathcal{H}_1$. This results from

$$\|u\|_0^2 = a(u, Gu), \quad u \in \mathcal{H}_1.$$

If therefore $a(u, \chi) = 0$ for all χ in the range of G, $u = 0$ follows. Thus

$$\Lambda = \inf \frac{(f, Gf)}{\|Gf\|_0^2},$$

where the infimum is now taken over all $f \neq 0$ in $\mathcal{H}_0$. Due to the coercivity of the bilinear form, $\Lambda > 0$. We first express Λ in terms of the norm of G and show that

$$\|G\| = \Lambda^{-1}. \tag{5.27}$$

By the representation above, the estimate

$$\|Gf\|_0^2 \leq \Lambda^{-1}(f, Gf) \leq \Lambda^{-1} \|f\|_0 \|Gf\|_0$$

and therefore the upper estimate $\|G\| \leq \Lambda^{-1}$ follow. As the expression

$$(f, Gg) = a(Gf, Gg)$$

defines an inner product on $\mathcal{H}_0$, the Cauchy-Schwarz inequality yields

$$(Gf, g) \leq (f, Gf)^{1/2} (g, Gg)^{1/2}.$$

Inserting $g = Gf$ one obtains

$$\|Gf\|_0^2 \leq \|G\|(f, Gf), \tag{5.28}$$

which implies the lower estimate $\Lambda^{-1} \leq \|G\|$ and proves (5.27).

Let (f_n) now be a sequence of elements in $\mathcal{H}_0$ with

$$\|f_n\|_0 = 1, \quad \lim_{n \to \infty} \|Gf_n\|_0 = \|G\|.$$

As, by equation (5.28),

$$\frac{\|Gf_n\|_0^2}{\|G\|} \leq (f_n, Gf_n) \leq \|G\| \|f_n\|_0^2,$$

one obtains from (5.27)

$$\lim_{n \to \infty} (f_n, Gf_n) = \Lambda^{-1} = \lim_{n \to \infty} \|Gf_n\|_0.$$

The relation

$$\|f_n - \Lambda G f_n\|_0^2 = \Lambda^2 \|G f_n\|_0^2 - 2\Lambda (f_n, G f_n) + \|f_n\|_0^2$$

yields therefore finally

$$\lim_{n \to \infty} \|f_n - \Lambda G f_n\|_0 = 0.$$

Thus Λ is an approximate eigenvalue and hence belongs to the spectrum.

Conversely, every point $\lambda > 0$ in the spectrum is an approximate eigenvalue. Therefore there exists, for every $\varepsilon > 0$, an $f \in \mathscr{H}_0$ of norm 1 with

$$\lambda^{-1} = \|G f + \lambda^{-1}(f - \lambda G f)\|_0 \leq \|G\| + \varepsilon$$

so that $\lambda^{-1} \leq \|G\| = \Lambda^{-1}$ or $\Lambda \leq \lambda$. Because all λ in the spectrum are positive as already stated in Theorem 5.1, this proves the proposition. □

The typical situation with molecular Hamiltonians as ours is that the spectrum splits into an essential spectrum with a greatest lower bound $\Sigma^* > \Lambda$ and a discrete spectrum then necessarily containing eigenvalues $\lambda < \Sigma^*$. In the case of the hydrogen atom, for example, the discrete spectrum consists of the eigenvalues

$$\lambda = -\frac{1}{2n^2}, \quad n = 1, 2, 3, \dots$$

that cluster at the minimum of the essential spectrum. This had been interpreted by Bohr as a quantum effect and was explained by Schrödinger in his seminal paper [73] in which he first stated his equation. The hydrogen eigenvalues are calculated in Sect. 9.4 and are depicted in Fig. 5.1. Our next theorems aim at such situations. They form the mathematical basis of the Ritz method to compute the eigenvalues corresponding to the ground state and the excited states of atoms and molecules.

Fig. 5.1 Discrete spectrum and minimum of the essential spectrum of the hydrogen atom

Theorem 5.8. *Let the interval $\lambda < \Sigma$ contain only points of the discrete spectrum, that is, isolated eigenvalues of finite multiplicity. Let the subspace $\mathscr{E}$ of $\mathscr{H}_0$ be invariant under G and contain the eigenvectors for all eigenvalues in this interval and let $\mathscr{E}^\perp$ be the orthogonal complement of $\mathscr{E}$ in $\mathscr{H}_0$. The value*

$$\lambda_* = \inf\left\{ a(u, u) \mid u \in \mathscr{H}_1 \cap \mathscr{E}^\perp, \|u\|_0 = 1 \right\} \tag{5.29}$$

belongs then itself to the spectrum and is greater than or equal Σ.

Proof. The vector spaces $\mathcal{E}^{\perp}$ and $\mathcal{H}_1 \cap \mathcal{E}^{\perp}$ are closed subspaces of $\mathcal{H}_0$ and $\mathcal{H}_1$, respectively, and therefore themselves Hilbert spaces to which our theory applies. As with $\mathcal{E}$ also $\mathcal{E}^{\perp}$ is invariant under G, the restriction of G to $\mathcal{E}^{\perp}$ plays then the role of G. By Theorem 5.7 and Theorem 5.4, applied to $\mathcal{E}^{\perp}$ and $\mathcal{H}_1 \cap \mathcal{E}^{\perp}$ in place of $\mathcal{H}_0$ and $\mathcal{H}_1$, there exist therefore elements $f_n \in \mathcal{E}^{\perp}$ for which

$$\lim_{n \to \infty} \|(I - \lambda_* G)f_n\|_0 = 0, \quad \|f_n\|_0 = 1. \tag{5.30}$$

The quantity λ_* thus belongs, by Theorem 5.4, to the spectrum in the original sense. We show that the assumption $\lambda_* < \Sigma$ leads to a contradiction. The reason is that, under the given assumptions, λ_* would then be an isolated eigenvalue of finite multiplicity and the associated eigenspace $\mathcal{E}_*$ a subspace of $\mathcal{E}$, or conversely $\mathcal{E}^{\perp}$ a subspace of $\mathcal{E}_*^{\perp}$. The f_n above would then belong to $\mathcal{E}_*^{\perp}$ so that, by (5.19),

$$\|f_n\|_0 \le \vartheta^{-1} \|(I - \lambda_* G)f_n\|_0$$

with a certain constant ϑ. But this contradicts (5.30). $\qquad\qquad\square$

Our considerations now culminate in the min-max principle on which the Rayleigh-Ritz variational method to compute the eigenvalues and eigenvectors is based.

Theorem 5.9. *Let $u_1, \dots, u_m$ be pairwise orthogonal normed eigenvectors for the isolated eigenvalues $\lambda_1 \le \dots \le \lambda_m$ of finite multiplicity. Let the interval $\lambda \le \lambda_m$ contain no other point of the spectrum and let, for $m \ge 2$, the vectors $u_1, \dots, u_{m-1}$ span the eigenspaces for the eigenvalues $\lambda < \lambda_m$. Then*

$$\lambda_m = \min_{\mathcal{V}_m} \max_{v \in \mathcal{V}_m} \frac{a(v, v)}{(v, v)}, \tag{5.31}$$

where the minimum is taken over all m-dimensional subspaces $\mathcal{V}_m$ of $\mathcal{H}_1$ and the maximum, without explicitly stating this every time, over all $v \ne 0$ in $\mathcal{V}_m$.

Proof. Let $\mathcal{E}_j$ be the subspace spanned by the vectors $u_1, \dots, u_j$. If $m = 1$, the proof starts from the observation that, by Theorem 5.7, $\lambda_1 = \Lambda$ is the infimum of the Rayleigh quotient and λ_1 therefore represents a lower bound for the right-hand side of (5.31). Choosing the subspace $\mathcal{V}_1 = \mathcal{E}_1$, one sees that the value λ_1 is attained.

If $m > 1$, λ_m is an upper bound for the right-hand side of (5.31), as one recognizes inserting $\mathcal{E}_m$ for $\mathcal{V}_m$. To prove that the maximum over an arbitrarily given m-dimensional subspace $\mathcal{V}_m$ is $\ge \lambda_m$ and λ_m therefore also a lower bound for the right-hand side of (5.31), fix a basis $v_1, \dots, v_m$ of $\mathcal{V}_m$. Let $a \in \mathbb{R}^m$, $a \ne 0$, be a vector that is orthogonal to the vectors $x_1, \dots, x_{m-1} \in \mathbb{R}^m$ with the components

$$x_k|_i = (v_i, u_k), \quad i = 1, \dots, m.$$

The vector

$$v^* = \sum_{i=1}^{m} a|_i v_i \in \mathcal{V}_m$$

satisfies then the orthogonality conditions

$$(v^*, u_k) = \sum_{i=1}^{m} (v_i, u_k) a|_i = x_k^T a = 0$$

for $k = 1, \ldots, m-1$. Therefore $v^* \in \mathscr{E}_{m-1}^{\perp}$ and

$$\max_{v \in \mathscr{V}_m} \frac{a(v,v)}{(v,v)} \geq \frac{a(v^*, v^*)}{(v^*, v^*)} \geq \min_{v \in \mathscr{E}_{m-1}^{\perp}} \frac{a(v,v)}{(v,v)}.$$

As $\mathscr{E}_{m-1}$ contains by assumption all eigenvectors for the eigenvalues λ below λ_m, Theorem 5.8 finally shows that the rightmost expression is $\geq \lambda_m$. □

The crucial point with the min-max principle is that no a priori information on the eigenvalues or eigenspaces is needed, which makes it an extremely powerful tool not only to give bounds for the eigenvalues but also to compute them.

5.3 The Rayleigh-Ritz Method

The Rayleigh-Ritz method is a variational method to compute the eigenvalues below the essential spectrum and the corresponding eigenvectors. It has the advantage of being based on minimal, very general assumptions and produces optimal solutions in terms of the approximation properties of the underlying trial spaces. We do not advocate the method as standard numerical procedure for the electronic Schrödinger equation but include this section to show how the approximation properties of finite dimensional subspaces transfer to the solution of the eigenvalue problem. The theory of the Rayleigh-Ritz method has to a large extent been developed in the context of finite element methods, see [8,9], or [68]. A recent convergence theory and a survey of the current literature can be found in [50].

We start from the same abstract framework as in the preceding section and from assumptions as in Theorem 5.9 in particular. Let $u_1, \ldots, u_m$ be pairwise orthogonal normed eigenvectors for the isolated eigenvalues $\lambda_1 \leq \ldots \leq \lambda_m$ of finite multiplicity. Let the interval $\lambda \leq \lambda_m$ contain no other point of the spectrum and let, in the case that $m \geq k \geq 2$, the vectors $u_1, \ldots, u_{k-1}$ span the eigenvectors for the eigenvalues $\lambda < \lambda_k$. As in the preceding section, let $\mathscr{E}_k$ denote the subspace spanned by $u_1, \ldots, u_k$.

Let $\mathscr{S}$ be a subspace of $\mathscr{H}_1$ of a dimension $n \geq m$. Then there exist pairwise orthogonal normed vectors $u'_1, \ldots, u'_n \in \mathscr{S}$ and real numbers $\lambda'_1, \ldots, \lambda'_n$ with

$$a(u'_k, v) = \lambda'_k (u'_k, v), \quad v \in \mathscr{S}. \tag{5.32}$$

Without restriction, let $\lambda'_1 \leq \ldots \leq \lambda'_n$. As will be shown, the quantities $\lambda'_1, \ldots, \lambda'_m$ approximate then the eigenvalues $\lambda_1, \ldots, \lambda_m$ of the original problem and the u'_k the corresponding eigenvectors in a sense explained later. This already fixes the method, which replicates the weak form of the eigenvalue problem and is completely determined by the choice of the subspace $\mathscr{S}$ replacing the original solution space.

Computationally, one starts from a basis $\varphi_1, \ldots, \varphi_n$ of $\mathscr{S}$ and calculates the symmetric and positive definite $(n \times n)$-matrices A and M with the entries

$$A|_{ij} = a(\varphi_i, \varphi_j), \quad M|_{ij} = (\varphi_i, \varphi_j). \tag{5.33}$$

If the discrete eigenvectors $u'_k \in \mathscr{S}$ have the representation

$$u'_k = \sum_{i=1}^{n} x_k|_i \, \varphi_i, \tag{5.34}$$

the coefficient vectors $x_k \in \mathbb{R}^n$ solve the algebraic eigenvalue problem

$$A x_k = \lambda'_k M x_k, \quad x_k^T M x_l = \delta_{kl}. \tag{5.35}$$

The existence of a complete set of M-orthogonal eigenvectors x_k follows from the spectral theorem of linear algebra. The relevant x_k and λ'_k can be computed by the standard methods of numerical linear algebra like the Lanczos method or, often the better choice in the present context, by preconditioned inverse iteration methods.

A first, but fundamental and very important observation on the relation between the original eigenvalues λ_k and their discrete counterparts λ'_k is:

Theorem 5.10. *Independent of the choice of the subspace $\mathscr{S}$, always*

$$\lambda_k \leq \lambda'_k, \quad k = 1, \ldots, m. \tag{5.36}$$

Proof. The proof is a simple consequence from the min-max principle. Let $\mathscr{V}_k$ be the k-dimensional subspace of $\mathscr{H}_1$ spanned by $u'_1, \ldots, u'_k$. Then

$$\lambda'_k = \max_{v \in \mathscr{V}_k} \frac{a(v, v)}{(v, v)},$$

from which the proposition follows with Theorem 5.9.   □

To give lower estimates and to bound the error, the approximation properties of the spaces $\mathscr{S}$ have to be brought into play. They are measured in terms of the a-orthogonal projection operator $P : \mathscr{H}_1 \to \mathscr{S}$ defined by

$$a(Pu, v) = a(u, v), \quad v \in \mathscr{S}. \tag{5.37}$$

With respect to the energy norm given by $\|v\|^2 = a(v, v)$, the projection Pu is the best approximation of $u \in \mathscr{H}_1$ by an element of $\mathscr{S}$, which means that for all $v \in \mathscr{S}$

$$\|u - Pu\| \leq \|u - v\|. \tag{5.38}$$

We remark that the approximation Pu of the solution $u \in \mathcal{H}_1$ of the equation

$$a(u,v) = f^*(v), \quad v \in \mathcal{H}_1, \tag{5.39}$$

with f^* a given bounded linear functional on $\mathcal{H}_1$ can be computed without the knowledge of u. In finite element methods, Pu is the approximate solution.

Theorem 5.11. *For $k = 1, \ldots, m$ given, let*

$$d_k = \sup \left\{ \|u - Pu\| \,\big|\, u \in \mathcal{E}_k, \|u\|_0 = 1 \right\} \tag{5.40}$$

denote the distance from $\mathcal{E}_k$ to $\mathcal{S}$. Provided that $d_k^2 \le \lambda_1/4$ then

$$0 \le \frac{\lambda_k' - \lambda_k}{\lambda_k} \le \frac{4}{\lambda_1} d_k^2. \tag{5.41}$$

Proof. We first introduce the constant

$$\sigma_k = \inf \left\{ \|Pu\|_0 \,\big|\, u \in \mathcal{E}_k, \|u\|_0 = 1 \right\}.$$

If we suppose for a moment that this σ_k is greater than zero, the subspace of $\mathcal{S}$ spanned by the vectors $Pu_1, \ldots, Pu_k$ has dimension k. By the min-max principle from Theorem 5.9, now applied to the restricted eigenvalue problem on $\mathcal{S}$,

$$\lambda_k' \le \max_{u \in \mathcal{E}_k} \frac{a(Pu, Pu)}{\|Pu\|_0^2}.$$

As Pu is the a-orthogonal projection of u onto $\mathcal{S}$, one further obtains

$$\lambda_k' \le \max_{u \in \mathcal{E}_k} \frac{a(u,u)}{\|Pu\|_0^2} \le \frac{1}{\sigma_k^2} \max_{u \in \mathcal{E}_k} \frac{a(u,u)}{(u,u)} = \frac{1}{\sigma_k^2} \lambda_k$$

by the definition of σ_k, or, using Theorem 5.10 for the lower estimate,

$$0 \le \frac{\lambda_k' - \lambda_k}{\lambda_k} \le \frac{1}{\sigma_k^2} - 1.$$

Therefore it remains to estimate σ_k in terms of the constant (5.40). Let

$$u = \sum_{i=1}^k \alpha_i u_i \in \mathcal{E}_k$$

be an arbitrary vector in $\mathcal{E}_k$ of norm $\|u\|_0 = 1$. Then

$$\|Pu\|_0^2 = \|u\|_0^2 - 2(u, u - Pu) + \|u - Pu\|_0^2 \ge 1 - 2(u, u - Pu).$$

Utilizing $(u_i, v) = \lambda_i^{-1} a(u_i, v)$ and $\lambda_i \geq \lambda_1 > 0$, one gets

$$|(u, u - Pu)| = \left| \sum_{i=1}^{k} \alpha_i \lambda_i^{-1} a(u_i, u - Pu) \right| \leq \lambda_1^{-1} \sum_{i=1}^{k} |\alpha_i a(u_i, u - Pu)|.$$

Choosing constants $\theta_i = \pm 1$ such that

$$\theta_i \alpha_i a(u_i, u - Pu) \geq 0$$

and introducing the new vector

$$v = \sum_{i=1}^{k} \theta_i \alpha_i u_i \in \mathscr{E}_k,$$

one obtains the estimate

$$|(u, u - Pu)| \leq \lambda_1^{-1} \sum_{i=1}^{k} \theta_i \alpha_i a(u_i, u - Pu) = \lambda_1^{-1} a(v, u - Pu).$$

Due to the symmetry of the bilinear form and the definition of P thus

$$|(u, u - Pu)| \leq \lambda_1^{-1} a(v - Pv, u - Pu) \leq \lambda_1^{-1} \|v - Pv\| \, \|u - Pu\|.$$

As also $\|v\|_0^2 = 1$, by the definition of d_k this implies

$$\left| (u, u - Pu) \right| \leq \frac{1}{\lambda_1} d_k^2.$$

Hence, passing to the infimum over the normed vectors $u \in \mathscr{E}_k$,

$$\sigma_k^2 \geq 1 - \frac{2}{\lambda_1} d_k^2.$$

Inserting this above and using $d_k^2 \leq \lambda_1/4$, the proposition follows.   □

The point is that the square of the distance (5.40) enters into the error estimate (5.41). The eigenvalues are thus much better approximated than is possible for the eigenvectors. For the minimum eigenvalue the estimate (5.41) reduces to

$$0 \leq \lambda_1' - \lambda_1 \leq 4 \|u_1 - Pu_1\|^2. \tag{5.42}$$

We now turn to the approximation of the eigenvectors. The problem here is that in general there is no unique correspondence between the original eigenvectors and their discretized counterparts and that a multiple eigenvalue can split into a cluster of discrete eigenvalues. The following theorem reflects this:

Theorem 5.12. *Let $u \in \mathscr{H}_1$ be an eigenvector for the eigenvalue λ. Then*

$$\left\| u - \sum_{|\mu'_k - \mu| < r} (u, u'_k) u'_k \right\|_0 \leq \frac{1}{r\lambda} \| u - Pu \|_0, \tag{5.43}$$

where $\mu = 1/\lambda$ and $\mu'_k = 1/\lambda'_k$ has been set and $0 < r \leq 1/\lambda$ is arbitrary.

Proof. We first represent the difference to be estimated in the form

$$u - \sum_{|\mu'_k - \mu| < r} (u, u'_k) u'_k = \sum_{|\mu'_k - \mu| \geq r} (u, u'_k) u'_k + u - \sum_{k=1}^{n} (u, u'_k) u'_k$$

and replace the inner products in the first sum on the right hand side by

$$(u, u'_k) = \frac{\mu}{\mu - \mu'_k} (u - Pu, u'_k).$$

This is possible as u is an eigenvector and the u'_k are discrete eigenvectors. With that

$$(u, u'_k) = \lambda^{-1} a(u, u'_k) = \lambda^{-1} a(u'_k, Pu) = \lambda^{-1} \lambda'_k (Pu, u'_k).$$

The resulting error representation reads in abbreviated form

$$u - \sum_{|\mu'_k - \mu| < r} (u, u'_k) u'_k = \frac{1}{\lambda} R(u - Pu) + (I - P_0)(u - Pu),$$

where the operator R and the $\mathscr{H}_0$-orthogonal projection P_0 onto $\mathscr{S}$ are given by

$$Rf = \sum_{|\mu'_k - \mu| \geq r} \frac{1}{\mu - \mu'_k} (f, u'_k) u'_k, \quad P_0 f = \sum_{k=1}^{n} (f, u'_k) u'_k.$$

Expressing the norms in terms of the expansion coefficients in the orthonormal basis of $\mathscr{S}$ consisting of the discrete eigenvectors $u'_1, \ldots, u'_n$, one finds

$$\| Rf \|_0^2 = \sum_{|\mu'_k - \mu| \geq r} \left| \frac{1}{\mu - \mu'_k} (f, u'_k) \right|^2 \leq \frac{1}{r^2} \| P_0 f \|_0^2.$$

This estimate is used to estimate the first term in the error representation. The proposition follows from the orthogonality properties of the different terms. □

The larger r is chosen, the more discrete eigenvectors u'_k are used to approximate the given eigenvector u and the smaller the error is, but the less specific the relation between the original and the discrete eigenvectors becomes. If the considered eigenvalue λ is sufficiently well separated from its neighbors λ', one can set

$$r = \frac{1}{2} \min_{\lambda' \neq \lambda} \left| \frac{1}{\lambda} - \frac{1}{\lambda'} \right|, \tag{5.44}$$

or $r = 1/\lambda$ should this lead to a value $r\lambda > 1$. As the approximate eigenvalues cluster around the exact ones, asymptotically then only approximate eigenvalues tending to λ are taken into account. The choice (5.44) for the parameter r results in the factor

$$\frac{1}{r\lambda} = 2 \max_{\lambda' \neq \lambda} \left| \frac{\lambda'}{\lambda' - \lambda} \right| \tag{5.45}$$

in front of the norm on the right hand side of the error estimate. The smaller it is, the better the given eigenvalue λ is separated from its neighbors. For an eigenvector u for the minimum eigenvalue λ_1, the error estimate transfers then to

$$\left\| u - \sum_{|\mu'_k - \mu_1| < r} (u, u'_k) u'_k \right\|_0 \leq 2 \frac{\lambda_2}{\lambda_2 - \lambda_1} \| u - Pu \|_0. \tag{5.46}$$

If the eigenvalue λ belongs to a cluster of closely neighboring eigenvalues, the parameter r should be chosen accordingly and (5.43) be interpreted as a result on the approximation by an element in the corresponding discrete invariant subspace.

The natural norm associated with the problem is the energy norm induced by the bilinear form. This error norm is considered in the following theorem which applies to eigenvectors for eigenvalues that are well separated from their neighbors:

Theorem 5.13. *Denoting by u' the given projection of the eigenvector u from Theorem 5.12 onto the chosen span of discrete eigenvectors,*

$$\| u - u' \|^2 \leq \lambda \| u - u' \|_0^2 + \max_{|\mu'_k - \mu| < r} |\lambda'_k - \lambda| \, \| u \|_0^2. \tag{5.47}$$

Proof. The estimate immediately follows from the relation

$$\| u - u' \|^2 = \lambda \| u - u' \|_0^2 + \sum_{|\mu'_k - \mu| < r} (\lambda'_k - \lambda)(u, u'_k)^2$$

that is shown by a straightforward computation.   □

If the parameter r is chosen sufficiently small the discrete eigenvalues λ'_k inside the selected interval tend asymptotically to λ. Combining the estimate from Theorem 5.13 with those from Theorem 5.11 and Theorem 5.12, one recognizes that the energy norm of the error tends to zero as fast as the energy norm distance (5.40) of the corresponding invariant subspace to the trial spaces. The Rayleigh-Ritz method in this respect fully exhibits the approximation properties of the trial spaces, however these are chosen, and is in this sense optimal.

Remarkably only the approximation error $\| u - Pu \|_0$ of the considered eigenvector u enters into the estimate (5.43). The estimate (5.43) differs in this respect from the error estimate (5.42) for the eigenvalues and the energy norm estimate (5.47) into which additionally the approximation error of all eigenvectors for eigenvalues below the considered one enters. To overcome this drawback, we assume for the rest of this section that the $\mathcal{H}_0$-orthogonal projection P_0 onto the ansatz space $\mathcal{S}$ is stable in the energy, or equivalently, the $\mathcal{H}_1$-norm, that is, that there is a κ with

$$\|P_0 v\| \leq \kappa \|v\|, \quad v \in \mathcal{H}_1. \tag{5.48}$$

The idea is that κ should be independent of hidden discretization parameters. This holds, for example, for certain spectral methods, for wavelets, and in the finite element case, there at least under some restrictions on the underlying grids [14, 17].

Theorem 5.14. *Let $u \in \mathcal{H}_1$ be an eigenvector for the eigenvalue λ. Then*

$$\left\| u - \sum_{|\mu'_k - \mu| < r} (u, u'_k) u'_k \right\| \leq \frac{2\kappa + 1}{r\lambda} \|u - Pu\|, \tag{5.49}$$

where $\mu = 1/\lambda$ and $\mu'_k = 1/\lambda'_k$ has been set and $0 < r \leq 1/\lambda$ is arbitrary.

Proof. The proof of (5.49) is based on the same error representation as that of Theorem 5.12 and transfers almost verbatim. Particularly it uses the norm estimate

$$\|Rf\|^2 = \sum_{|\mu'_k - \mu| \geq r} \lambda'_k \left| \frac{1}{\mu - \mu'_k} (f, u'_k) \right|^2 \leq \frac{1}{r^2} \|P_0 f\|^2.$$

The only exception is that in the final step one can no longer argue using the orthogonality properties of the different terms but has to switch to the triangle inequality. At this point the bound for the norm of the operator P_0 enters in form of the estimate

$$\|P_0(u - Pu)\| \leq \kappa \|u - Pu\|$$

for the projection of the approximation error. □

It is not astonishing that a similar error estimate holds for the higher eigenvalues, at least for those that are sufficiently well separated from the eigenvalues below them:

Theorem 5.15. *Let $u \in \mathcal{H}_1$ be a normed eigenvector for the eigenvalue λ. Assume that $\lambda'_k \geq \lambda$ for all discrete eigenvalues λ'_k in the neighborhood of λ fixed by the condition $|\mu'_k - \mu| < r$, where again $\mu = 1/\lambda$, $\mu'_k = 1/\lambda'_k$, and $0 < r \leq 1/\lambda$. Then*

$$\min_{\lambda'_k \geq \lambda} (\lambda'_k - \lambda) \leq \left(\frac{2\kappa + 1}{r\lambda} \right)^2 \|u - Pu\|^2, \tag{5.50}$$

provided that there is already a discrete eigenvalue $\lambda'_k \geq \lambda$ for which $\lambda'_k - \lambda \leq \lambda$.

Proof. Denoting by u' the given projection of u from Theorem 5.12 or Theorem 5.14 onto the chosen span of discrete eigenvectors, as in the proof of Theorem 5.13

$$\|u - u'\|^2 = \lambda \|u - u'\|_0^2 + \sum_{|\mu'_k - \mu| < r} (\lambda'_k - \lambda)(u, u'_k)^2.$$

Since the given differences $\lambda'_k - \lambda$ are by assumption nonnegative, this implies

$$\|u - u'\|^2 \geq \lambda \|u - u'\|_0^2 + \min_{\lambda'_k \geq \lambda} (\lambda'_k - \lambda) \|u'\|_0^2.$$

Since u' and $u - u'$ are by definition $\mathcal{H}_0$-orthogonal and $\|u\|_0 = 1$, this means

$$\|u - u'\|^2 \geq \min_{\lambda'_k \geq \lambda} (\lambda'_k - \lambda) + \lambda \left(1 - \min_{\lambda'_k \geq \lambda} \frac{\lambda'_k - \lambda}{\lambda}\right) \|u - u'\|_0^2.$$

As the second term on the right-hand side of this inequality is by assumption non-negative, the proposition follows from Theorem 5.14. $\square$

We remark that one can even get rid of the assumption that there is already a discrete eigenvalue $\lambda'_k \geq \lambda$ for which $\lambda'_k - \lambda \leq \lambda$ at the price of a slightly more complicated expression on the right hand side of the error estimate. If there is a discrete eigenvalue $\lambda'_k < \lambda$, the best possible choice for the parameter r is given by

$$\frac{1}{r\lambda} = \max\left\{1, \max_{\lambda'_k < \lambda} \frac{\lambda'_k}{\lambda - \lambda'_k}\right\}. \tag{5.51}$$

Assuming the energy norm stability (5.48) of the $\mathcal{H}_0$-orthogonal projection onto the ansatz space, the method can thus take full advantage of a higher regularity of the considered eigenvector or eigenfunction compared to the other ones, particularly compared to those for lower eigenvalues. It should further be noted that in the finite-element context one gains, depending on the regularity of the problem, up to one order of approximation in the $\mathcal{H}_0$-norm compared to the $\mathcal{H}_1$-norm. By Theorem 5.12 this property transfers to the approximate eigenfunctions.

5.4 The Lower Bound of the Essential Spectrum

We return in this section to the electronic Schrödinger equation, that is, the bilinear form introduced in Sect. 4.1. The results of the previous two sections transfer to this case if one replaces the given bilinear form by a shifted variant as in (4.16). We recall the definition (4.1) of the minimum energy $\Lambda(\sigma)$ and of the ionization threshold $\Sigma(\sigma)$ from Sect. 5.1. The aim of this section is to translate our basic assumption (5.4) on these two quantities into a statement about the spectrum.

We begin with an intermediate result that holds for much more general cases than only for the electronic Schrödinger equation, for example for Schrödinger operators with locally integrable potentials that are bounded from below. If necessary, the solution space $H^1(\sigma)$ has then to be replaced by a corresponding subspace.

Lemma 5.2. *For all λ in the essential spectrum and all $R > 0$,*

$$\lambda \geq \Sigma(R, \sigma), \tag{5.52}$$

that is, the $\Sigma(R, \sigma)$ remain bounded if the essential spectrum is non-empty.

Proof. The proof relies on the fact that there exists, for every $R > 0$, an infinitely differentiable function η that depends on R, has a compact support, and for which

$$a(u, u) + (\eta u, u) \geq \Sigma(R, \sigma) \|u\|_0^2 \tag{5.53}$$

holds for all functions u in $\mathscr{D}(\sigma)$ and with that also in the solution space $H^1(\sigma)$.

To construct η, let $\phi_1, \phi_2 : \mathbb{R} \to [0, 1]$ be a pair of infinitely differentiable functions such that $\phi_1(r) = 0$ for $r \leq R$ and $\phi_1(r) = 1$ for $r \geq R+1$ and such that $\phi_1^2 + \phi_2^2 = 1$ everywhere. Let $\chi_1(x) = \phi_1(|x|)$ and $\chi_2(x) = \phi_2(|x|)$. Then

$$\chi_1(x)^2 + \chi_2(x)^2 = 1$$

for all $x \in \mathbb{R}^{3N}$. This implies

$$|\nabla u|^2 = |\nabla(\chi_1 u)|^2 + |\nabla(\chi_2 u)|^2 - \left(|\nabla \chi_1|^2 + |\nabla \chi_2|^2 \right) u^2$$

for all infinitely differentiable functions u with compact support. Thus

$$a(u, u) = a(\chi_1 u, \chi_1 u) + a(\chi_2 u, \chi_2 u) - \int \left(|\nabla \chi_1|^2 + |\nabla \chi_2|^2 \right) u^2 \, dx.$$

Since the bilinear form (4.13) satisfies for sufficiently large μ the estimate (4.16),

$$a(u, u) \geq a(\chi_1 u, \chi_1 u) - \int \left(\mu \chi_2^2 + |\nabla \chi_1|^2 + |\nabla \chi_2|^2 \right) u^2 \, dx$$

follows. If $u \in \mathscr{D}(\sigma)$, also $\chi_1 u \in \mathscr{D}(\sigma)$. Since $\chi_1(x) = 0$ for $|x| \leq R$ therefore

$$a(\chi_1 u, \chi_1 u) \geq \Sigma(R, \sigma) \|\chi_1 u\|_0^2$$

by the definition (5.2) of the constant $\Sigma(R, \sigma)$. Because

$$\|\chi_1 u\|_0^2 = \|u\|_0^2 - \int \chi_2^2 u^2 \, dx,$$

this proves the estimate (5.53) with the infinitely differentiable function

$$\eta(x) = (\Sigma(R, \sigma) + \mu) \phi_2(r)^2 + \phi_1'(r)^2 + \phi_2'(r)^2$$

vanishing for $r \geq R+1$, where $|x| = r$ has been set.

The second main ingredient of the proof is the fact that every H^1-bounded sequence of functions possesses a subsequence that converges on every bounded set in the L_2-sense. Let λ now be a point in the essential spectrum. By the corollary from Theorem 5.6 there exists then a sequence of functions $u_n \in H^1(\sigma)$ with

$$\|u_n\|_0 = 1, \quad u_n \to 0 \quad \text{weakly in } L_2(\sigma),$$

$$\lim_{n \to \infty} a(u_n, u_n) = \lambda.$$

The estimate (4.16) shows that then also a joint bound for the H^1-norms of these functions exists. We can thus additionally assume that the u_n converge in the L_2-sense to a limit function u^* on the bounded support of the function η from (5.53). But as the functions u_n converge weakly to zero in L_2, necessarily $u^* = 0$. Hence

$$\lim_{n \to \infty} (\eta u_n, u_n) = 0.$$

As, by (5.53) and because $\|u_n\|_0 = 1$, for all n

$$a(u_n, u_n) + (\eta u_n, u_n) \geq \Sigma(R, \sigma),$$

one obtains in the limit the upper bound $\lambda \geq \Sigma(R, \sigma)$ for the the constants (5.2) or, the other way around, a lower bound for the essential spectrum. $\square$

In other words, if the essential spectrum is non-empty the limit

$$\Sigma(\sigma) = \lim_{R \to \infty} \Sigma(R, \sigma) \tag{5.54}$$

remains finite and forms a lower bound of the essential spectrum. Conversely, if the $\Sigma(R, \sigma)$ tend to infinity, the essential spectrum is empty. The subspace spanned by the eigenfunctions for the eigenvalues in the discrete spectrum is then dense in the solution space as can be seen applying Theorem 5.7 to its orthogonal complement.

The next lemma shows that the limit (5.54) is, if finite, not only a lower bound for the essential spectrum but in fact its greatest lower bound, its infimum:

Lemma 5.3. *If there is no point $\lambda \leq \Sigma$ in the essential spectrum, then for all $\varepsilon > 0$,*

$$\Sigma - \varepsilon \leq \Sigma(R, \sigma) \tag{5.55}$$

for all R that are sufficiently large in dependence of ε.

Proof. By Theorem 5.5, all accumulation points of eigenvalues belong to the essential spectrum. By Theorem 5.7, the interval $\lambda < \Lambda(\sigma)$ is a subset of the resolvent. The interval $\lambda \leq \Sigma$ can thus contain at most finitely many eigenvalues of finite multiplicity and no other point in the spectrum. If it does not contain a point of the spectrum, the proposition follows from Theorem 5.7. Otherwise, let the L_2-orthogonal normed eigenfunctions $u_1, \ldots, u_n$ span the corresponding eigenspaces and let

$$Pu = \sum_{k=1}^{n} (u, u_k) u_k$$

denote the L_2- and a-orthogonal projection onto the subspace spanned by these eigenfunctions. For all functions $u \in \mathscr{D}(\sigma)$ by Theorem 5.8 then

$$a(u - Pu, u - Pu) \geq \Sigma \|u - Pu\|_0^2.$$

A short calculation shows

$$a(u, u) = a(u - Pu, u - Pu) + \sum_{k=1}^{n} \lambda_k (u, u_k)^2,$$

$$\|u - Pu\|_0^2 = \|u\|_0^2 - \sum_{k=1}^{n} (u, u_k)^2.$$

With help of the relation above one concludes that

$$a(u, u) \geq \Sigma \|u\|_0^2 - \sum_{k=1}^{n} (\Sigma - \lambda_k)(u, u_k)^2$$

holds for all functions $u \in \mathscr{D}(\sigma)$ and particularly for those that vanish on the ball of radius R around the origin and have L_2-norm 1. Taking the infimum over all these u

$$\Sigma(R, \sigma) \geq \Sigma - \sum_{k=1}^{n} (\Sigma - \lambda_k) \|\chi_R u_k\|_0^2$$

follows, where χ_R denotes the characteristic function of the exterior of the ball of radius R around the origin. Since the L_2-norm of the functions the $\chi_R u_k$ tends to zero as R tends to infinity, the proposition follows choosing R sufficiently large. $\square$

Like the previous lemma, this lemma holds for much more general cases than only the electronic Schrödinger equation, particularly for Schrödinger operators with locally integrable potentials. We can conclude that the essential spectrum is empty if and only if the constants (5.2) tend to infinity as R tends to infinity. In this case, the linear combinations of the eigenfunctions are dense in the given Hilbert space. Every function in this space can be expanded into these eigenfunctions. If the limit (5.54) remains finite, the essential spectrum is non-empty and the ionization threshold $\Sigma(\sigma)$ is not only its greatest lower bound but even its minimum, since it is an accumulation point of the essential spectrum. Remembering Lemma 5.1 we obtain:

Theorem 5.16. *The essential spectrum of the electronic Schrödinger operator is non-empty. Its minimum is the ionization threshold $\Sigma(\sigma) \leq 0$ from (5.3). The minimum energy $\Lambda(\sigma) < \Sigma(\sigma)$ from (5.1) is an isolated eigenvalue of finite multiplicity.*

The eigenfunctions for the eigenvalue $\lambda = \Lambda(\sigma)$ are the ground states of the system with the spin distribution kept fixed, and the minimum eigenvalue $\Lambda(\sigma)$ itself the ground state energy. The greatest lower bound of the essential spectrum for the full, spin-dependent problem can be determined with the same techniques. It is equal the minimum of the bounds $\Sigma(\sigma)$ obtained for the components.

The information that Theorem 5.16 provides is by far not all what is known for Hamilton operators of atoms and molecules. Important results are the Hunziker-van Winter-Zhislin theorem [46, 90, 98] that characterizes the ionization threshold as the energy threshold above which such a system can break apart, or the fact that atoms and positively charged ions have an infinite discrete spectrum below the ionization threshold. We refer to the survey article [47] or monographs on mathematical physics like [38, 71], or [88] for an in-depth discussion of such topics.

5.5   The Exponential Decay of the Eigenfunctions

The spectral properties of Schrödinger operators are strongly intertwined with the exponential decay of their eigenfunctions for eigenvalues below the essential spectrum. The first results of this type for more than three electrons are due to Ahlrichs [4] for the case of a single nucleus, that is, an atom, and to O'Connor [20], who treated the general case and derived an isotropic L_2-bound. O'Connor's result was a short time after improved by Combes and Thomas [19]. Simon [74] found a pointwise isotropic bound. The actual decay behavior of the eigenfunctions is complicated and in general highly anisotropic. A first result in this direction was proven by Deift, Hunziker, Simon, and Vock [22]. In some sense the final study is Agmon's monograph [3]. Agmon introduced the Agmon distance, named after him, with the help of which the decay of the eigenfunctions can be described rather precisely.

The isotropic L_2-decay of the eigenfunctions plays a central role for this work because we want to show, on the basis of this result, that also many of the high-order mixed derivatives of the eigenfunctions decay exponentially. For this reason, and to keep the presentation as self-contained as possible, we give a short proof of O'Connor's theorem that closely follows Agmon's argumentation [3]. It starts directly from the definition (5.3) of the ionization threshold and does not utilize the fact that it represents the infimum of the essential spectrum.

Theorem 5.17. *Let $\lambda < \Sigma(\sigma)$ be an eigenvalue below the ionization threshold (5.3) and $u \in H^1(\sigma)$ be an assigned eigenfunction. For $\lambda < \Sigma < \Sigma(\sigma)$, the functions*

$$x \;\to\; \exp\left(\sqrt{2(\Sigma-\lambda)}\,|x|\right) u(x), \;\; \exp\left(\sqrt{2(\Sigma-\lambda)}\,|x|\right)(\nabla u)(x) \qquad (5.56)$$

are then square integrable, that is, u and ∇u decay exponentially in the L_2-sense.

Proof. We begin choosing a radius R such that

$$\Sigma(R,\sigma) - \Sigma =: \alpha > 0. \qquad (5.57)$$

We further introduce the bounded functions

$$\delta(x) = \sqrt{2(\Sigma - \lambda)} \frac{|x|}{1 + \varepsilon|x|},$$

with $\varepsilon > 0$ given arbitrarily, and observe that

$$|(\nabla\delta)(x)|^2 \leq 2(\Sigma - \lambda)$$

for all $x \neq 0$ independent of the choice of ε. Since

$$\nabla(e^{-\delta}v) \cdot \nabla(e^{\delta}v) = \nabla v \cdot \nabla v - |\nabla\delta|^2 v^2,$$

this leads to the estimate

$$a(e^{-\delta}v, e^{\delta}v) \geq a(v,v) - (\Sigma - \lambda)\|v\|_0^2 \tag{5.58}$$

for all infinitely differentiable functions v that have a compact support and that vanish on a neighborhood of the origin. In particular, the estimate holds for the functions $v \in \mathscr{D}(\sigma)$ that takes the value 0 on the ball of radius R around the origin. For these v,

$$a(v,v) \geq \Sigma(R,\sigma)\|v\|_0^2.$$

In combination with (5.57) and (5.58), this yields

$$\alpha\|v\|_0^2 \leq a(e^{-\delta}v, e^{\delta}v) - \lambda\|v\|_0^2. \tag{5.59}$$

Next, we fix a rotationally symmetric, infinitely differentiable function χ that vanishes on the ball of radius R around the origin and takes the value $\chi(x) = 1$ for $|x| \geq R+1$. Let u in $\mathscr{D}(\sigma)$ be arbitrary. Setting $v = \chi e^{\delta}u$, (5.59) becomes

$$\alpha\|\chi e^{\delta}u\|_0^2 \leq a(\chi u, \chi e^{2\delta}u) - \lambda(\chi u, \chi e^{2\delta}u). \tag{5.60}$$

To shift the factor χ to the right hand side, we introduce the function

$$\eta = \frac{2\chi\nabla\chi \cdot \nabla\delta + |\nabla\chi|^2}{2}$$

that takes the value $\eta(x) = 0$ for $|x| \leq R$ and $|x| \geq R+1$. With help of the relation

$$\nabla(\chi u) \cdot \nabla(\chi e^{2\delta}u) = \nabla u \cdot \nabla(\chi^2 e^{2\delta}u) + 2\eta e^{2\delta}u^2,$$

the estimate (5.60) for the functions $u \in \mathscr{D}(\sigma)$ can then be rewritten as

$$\alpha\|\chi e^{\delta}u\|_0^2 \leq a(u, \chi^2 e^{2\delta}u) - \lambda(u, \chi^2 e^{2\delta}u) + (u, \eta e^{2\delta}u). \tag{5.61}$$

As χe^{δ}, the first-order derivatives of χe^{δ}, and $\eta e^{2\delta}$ are bounded and as $\mathcal{D}(\sigma)$ is a dense subspace of $H^1(\sigma)$, the estimate transfers to arbitrary functions $u \in H^1(\sigma)$.

Since $\chi^2 e^{2\delta} u \in H^1(\sigma)$, the first two terms on the right hand side of (5.61) cancel for the given eigenfunction u for the eigenvalue λ. The estimate thus reduces to

$$\alpha \|\chi e^{\delta} u\|_0^2 \leq (u, \eta e^{2\delta} u) \tag{5.62}$$

for this u. To estimate the H^1-norm of $\chi e^{\delta} u$, we recall that, by (4.16) and (5.58),

$$\frac{1}{4}\|v\|_1^2 \leq a(v,v) + \mu \|v\|_0^2 \leq a(e^{-\delta}v, e^{\delta}v) + (\Sigma - \lambda + \mu)\|v\|_0^2$$

for all infinitely differentiable functions v that have a compact support and that vanish on a neighborhood of the origin, where the constant $\mu > 0$ was more precisely specified in Sect. 4.1. From that one obtains, in the same way as above, the estimate

$$\frac{1}{4}\|\chi e^{\delta} u\|_1^2 \leq (u, \eta e^{2\delta} u) + (\Sigma + \mu)\|\chi e^{\delta} u\|_0^2$$

for the given eigenfunction u and, with (5.62), finally the estimate

$$\|\chi e^{\delta} u\|_1^2 \leq \left(4 + 4\frac{\Sigma + \mu}{\alpha}\right)(u, \eta e^{2\delta} u).$$

Since the functions $\eta e^{2\delta}$ and $\nabla \delta$ are uniformly bounded in ε, the L_2-norms of the functions $e^{\delta} u$ and $e^{\delta} \nabla u$ therefore remain bounded uniformly in ε. The proposition follows with the monotone convergence theorem letting ε tend to zero. □

The given decay rates cannot be improved without further assumptions on the considered system. This can already be recognized by the case of a single electron that moves in the field of a nucleus of charge Z, that is, by the Hamilton operator

$$H = -\frac{1}{2}\Delta - \frac{Z}{|x|}. \tag{5.63}$$

In this case, the ionization threshold and with that the bottom of the essential spectrum is $\Sigma^* = 0$. The ground state wave function and the associated eigenvalue are

$$u(x) = e^{-Z|x|}, \quad \lambda = -\frac{1}{2}Z^2, \tag{5.64}$$

up to normalization of u. For this example,

$$\exp\left(\sqrt{2(\Sigma^* - \lambda)}\,|x|\right) u(x) = 1 \tag{5.65}$$

so that the functions (5.56) cannot be square integrable for $\Sigma \geq \Sigma^*$. The same applies for the higher eigenfunctions of the operator (5.63), which can be found in almost every textbook on quantum mechanics and which are calculated in Chap. 9. Figure 5.2 shows a cross section through the exponentially decaying, rotationally symmetric ground state eigenfunction (5.64). Its singularity at the origin is typical for the behavior of electronic wave functions in the vicinity of the nuclei.

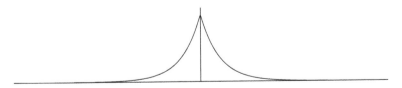

Fig. 5.2 The rotationally symmetric ground state eigenfunction of the hydrogen atom

The given technique of proof is by far not restricted to the electronic Schrödinger equation and can, for example, be applied to any Schrödinger operator

$$H = -\Delta + V, \tag{5.66}$$

with a locally integrable potential $V(x)$ tending to infinity as $|x|$ goes to infinity. Our considerations show that the essential spectrum of such operators is empty, that their eigenvalues tend to infinity, and that all their eigenfunctions tend in the L_2-sense faster to zero than any exponential function $x \to e^{\gamma|x|}$, $\gamma > 0$, grows.

Chapter 6
Existence and Decay of Mixed Derivatives

A primary aim of this work, and the decisive step to our analysis of the complexity of electronic wave functions, is to study the regularity of these functions. We want to show that they possess certain high-order square integrable weak derivatives and that these derivatives even decay exponentially, in the same way as the wave functions themselves. This goal is reached in the present chapter. A central idea of the proof is to examine instead of the solutions of the original Schrödinger equation the solutions of a modified equation for the correspondingly exponentially weighted wave functions. This equation is set up in the first section of this chapter and is based on the result on the exponential decay of the wave functions from Sect. 5.5. The study of the regularity in isotropic Hölder spaces in [32] is based on a similar idea. In Sect. 6.2 we introduce the high-order solution spaces and the corresponding norms. The actual proof relies on a mixture of variational techniques and Fourier analysis. The key is the estimates for the arising low-order terms, particularly for the nucleus-electron and the electron-electron interaction potential. These estimates are proven in Sect. 6.3 and Sect. 6.4. The estimates for the nucleus-electron interaction potential and an additional term coming from the exponential weights are in the end based on the Hardy inequality from Sect. 4.1, whose central role is reflected here again. In contrast to these estimates the estimates for the electron-electron interaction potential require that the considered functions satisfy the Pauli principle, that is, are antisymmetric with respect to the exchange of the positions of electrons with the same spin. The reason is that such functions vanish at the places where electrons with the same spin meet, which counterbalances the singularities of the electron-electron interaction potential. To derive these estimates and to master the arising singularities a further three-dimensional Hardy-type estimate is needed that holds only for functions vanishing at the origin. In Sect. 6.5 the regularity theorem for the exponentially weighted wave functions is stated and proven. This result serves then to derive bounds for the exponential decay of the mixed derivatives of the original wave functions. The present chapter is partly based on two former papers [92,94] of the author in which the existence of the mixed derivatives has been proven and estimates for their L_2-norms were given. The result on the exponential decay of these derivatives [95] was up to now only available on the author's website.

H. Yserentant, *Regularity and Approximability of Electronic Wave Functions*,
Lecture Notes in Mathematics 2000, DOI 10.1007/978-3-642-12248-4_6,
© Springer-Verlag Berlin Heidelberg 2010

6.1   A Modified Eigenvalue Problem

First we replace the rotationally symmetric exponential weight functions in (5.56) by products of weight functions that depend only on the coordinates of one single electron. Such weights are easier to analyze and fit into the framework that we will develop in the following sections. Let $u \in H^1(\sigma)$ be an eigenfunction for the eigenvalue $\lambda < \Sigma(\sigma)$. Let $\theta_1, \ldots, \theta_N \geq 0$ be given weight factors and let

$$F(x) = \gamma \sum_{i=1}^{N} \theta_i |x_i|, \quad \sum_{i=1}^{N} \theta_i^2 = 1. \tag{6.1}$$

Let γ be a decay rate as in Theorem 5.17, that is,

$$\gamma < \sqrt{2(\Sigma(\sigma) - \lambda)}, \tag{6.2}$$

and define the correspondingly exponentially weighted eigenfunction as

$$\tilde{u}(x) = \exp(F(x)) u(x). \tag{6.3}$$

This exponentially weighted eigenfunction solves then an eigenvalue equation that is similar to the original one. To derive it we start from the following two lemmata:

Lemma 6.1. *Let the function $u \in H^1$ and the constant $\gamma \in \mathbb{R}$ be first arbitrary. The function $\tilde{u}$ defined as in (6.3) is then not only locally square integrable but has also locally square integrable first-order weak partial derivatives. They read*

$$D_k \tilde{u} = e^F D_k F u + e^F D_k u, \tag{6.4}$$

where the operator D_k denotes weak differentiation for u and pointwise for F.

Proof. We first consider functions $u \in \mathscr{D}$, that is, infinitely differentiable functions with bounded support, and replace the function (6.1) by its smooth counterparts

$$F_\varepsilon(x) = \gamma \sum_{i=1}^{N} \theta_i \sqrt{|x_i|^2 + \varepsilon^2}. \tag{6.5}$$

Integration by parts then yields, for all test functions φ of the same type,

$$\int \left(e^{F_\varepsilon} D_k F_\varepsilon u + e^{F_\varepsilon} D_k u \right) \varphi \, dx = \int D_k \left(e^{F_\varepsilon} u \right) \varphi \, dx = - \int e^{F_\varepsilon} u \, D_k \varphi \, dx.$$

Letting ε tend to zero, one obtains, from the dominated convergence theorem,

$$\int \left(e^F D_k F u + e^F D_k u \right) \varphi \, dx = - \int e^F u \, D_k \varphi \, dx.$$

Since F and its first-order partial derivatives are bounded on the support of φ and $\mathscr{D}$ is a dense subspace of H^1, this relation transfers to all $u \in H^1$. This proves the differentiation formula above and transfers the product rule to the given case. $\qquad\square$

Lemma 6.2. *For all functions $u \in H^1$ and all test functions $v \in \mathscr{D}$,*

$$a\big(u, e^F v\big) - a\big(e^F u, v\big) = c\big(e^F u, v\big), \tag{6.6}$$

where $c(u,v)$ denotes the H^1-bounded bilinear form

$$c(u,v) = \frac{1}{2} \int \big\{ 2\nabla F \cdot \nabla u + \big(\Delta F - |\nabla F|^2\big) u \big\} v \, dx. \tag{6.7}$$

Proof. We consider again first only functions $u \in \mathscr{D}$ and replace F by its infinitely differentiable counterparts (6.5). A short calculation yields

$$\Delta\big(e^{F_\varepsilon} u\big) - e^{F_\varepsilon} \Delta u = 2\nabla F_\varepsilon \cdot \nabla\big(e^{F_\varepsilon} u\big) + \big(\Delta F_\varepsilon - |\nabla F_\varepsilon|^2\big) e^{F_\varepsilon} u.$$

If one multiplies this equation with a test function $v \in \mathscr{D}$ and integrates by parts

$$\int \nabla u \cdot \nabla\big(e^{F_\varepsilon} v\big) \, dx - \int \nabla\big(e^{F_\varepsilon} u\big) \cdot \nabla v \, dx$$

$$= \int \big\{ 2\nabla F_\varepsilon \cdot \nabla\big(e^{F_\varepsilon} u\big) + \big(\Delta F_\varepsilon - |\nabla F_\varepsilon|^2\big) e^{F_\varepsilon} u \big\} v \, dx$$

follows. As F_ε and ∇F_ε are locally uniformly bounded in $\varepsilon \leq \varepsilon_0$ and $|\Delta_i F_\varepsilon| \lesssim 1/|x_i|$, one can let ε tend to zero in this expression and recognizes with help of the dominated convergence theorem that (6.6) holds for all functions u and v in $\mathscr{D}$. The H^1-boundedness of the bilinear form (6.7) follows from the Hardy inequality. As the functions in $\mathscr{D}$ have a bounded support, both sides of equation (6.6) thus represent, by Lemma 6.1, bounded linear functionals in $u \in H^1$ for $v \in \mathscr{D}$ given. The equation transfers therefore to all functions $u \in H^1$ and all test functions $v \in \mathscr{D}$. $\qquad\square$

After these preparations we can now return to the initially introduced eigenfunction $u \in H^1(\sigma)$ for the eigenvalue λ and its exponentially weighted counterpart (6.3).

Theorem 6.1. *The exponentially weighted eigenfunction $\tilde{u}$ defined by (6.3) is itself contained in the space H^1 and solves the eigenvalue equation*

$$a(\tilde{u}, v) + \gamma s(\tilde{u}, v) = \tilde{\lambda}(\tilde{u}, v), \quad v \in H^1, \tag{6.8}$$

where the expression $s(u,v)$ denotes the H^1-bounded bilinear form

$$s(u,v) = \sum_{i=1}^{N} \theta_i \int \left\{ \frac{x_i}{|x_i|} \cdot \nabla_i u + \frac{1}{|x_i|} u \right\} v \, dx \tag{6.9}$$

and the real eigenvalue $\widetilde{\lambda} < \Sigma(\sigma) \leq 0$ is given by

$$\widetilde{\lambda} = \lambda + \frac{1}{2}\gamma^2. \tag{6.10}$$

Proof. The function (6.1) satisfies the estimate $F(x) \leq \gamma|x|$. Under the condition (6.2) the exponentially weighted eigenfunction (6.3) is therefore, by Theorem 5.17 and Lemma 6.1, contained in H^1. Setting $\widetilde{v} = e^F v$, by Lemma 6.1 and Lemma 6.2

$$a(\widetilde{u},v) + c(\widetilde{u},v) = a(u,\widetilde{v}) = \lambda(u,\widetilde{v}) = \lambda(\widetilde{u},v)$$

for all test functions $v \in \mathscr{D}$ and hence for all $v \in H^1$. The proposition follows calculating ∇F and ΔF explicitly and observing that $|\nabla F|^2 = \gamma^2$. $\qquad\square$

The next sections are devoted to the study of the modified eigenvalue problem (6.8) that the exponentially weighted eigenfunctions (6.3) satisfy. Hereby we take up a slightly more general approach and relax the symmetry properties prescribed by the Pauli principle a little bit. Let I be a nonempty subset of the set of the electron indices $1,\ldots,N$. Let $\mathscr{D}_I$ denote the subspace of $\mathscr{D}$ that consists of those functions in $\mathscr{D}$ that change their sign under the exchange of the electron positions x_i and x_j in $\mathbb{R}^3$ for indices $i \neq j$ in I. The closure of the subspace $\mathscr{D}_I$ in H^1 is the Hilbert space H_I^1. Our modified eigenvalue problem then consists in finding functions $u \neq 0$ in H_I^1 and values $\lambda < 0$ that satisfy the condition

$$a(u,v) + \gamma s(u,v) = \lambda(u,v), \quad v \in H_I^1. \tag{6.11}$$

Our aim is to study the regularity of the solutions of this eigenvalue problem in Hilbert spaces of mixed derivatives. Conditions on the parameter γ enter only implicitly since, with u a solution of (6.11) and with that also of equation (6.12) below, $\widetilde{u} = e^{-F}u$ is conversely a solution of the original eigenvalue equation (4.17) for which $e^F\widetilde{u}$ is then a square integrable function. We assume $\gamma \geq 0$ in the sequel.

Theorem 6.2. *Provided that the function (6.1) is symmetric with respect to the permutations of the electrons with indices $i \in I$, which is the case if and only if all θ_i for $i \in I$ are equal, a function $u \in H_I^1$ that solves (6.11) also solves the full equation*

$$a(u,v) + \gamma s(u,v) = \lambda(u,v), \quad v \in H^1. \tag{6.12}$$

That is, (6.11) does not only hold for test functions $v \in H_I^1$, but for all $v \in H^1$.

Proof. The proof is based on the observation that the affected bilinear forms are invariant under the considered permutations of the electrons, that is, on the fact that

$$a(u(P\cdot),v(P\cdot)) = a(u,v), \quad s(u(P\cdot),v(P\cdot)) = s(u,v)$$

for these permutations P, which follows from the invariance of the potential (4.9) and the function (6.1) under these permutations. Let G denote the group of permutations that fix the indices in the complement of I and define the operator

$$(\mathscr{A}v)(x) = \frac{1}{|G|} \sum_{P \in G} \text{sign}(P)v(Px),$$

that reproduces functions in $\mathscr{D}_I$ and H_I^1, respectively, and maps functions in H^1 to partially antisymmetric functions in H_I^1. Since, for arbitrary functions $u, v \in H^1$,

$$a(\mathscr{A}u, v) = a(u, \mathscr{A}v), \quad s(\mathscr{A}u, v) = s(u, \mathscr{A}v), \quad (\mathscr{A}u, v) = (u, \mathscr{A}v),$$

a solution $u \in H_I^1$ of (6.11) satisfies the equation

$$a(u,v) + \gamma s(u,v) = a(\mathscr{A}u, v) + \gamma s(\mathscr{A}u, v) = a(u, \mathscr{A}v) + \gamma s(u, \mathscr{A}v)$$
$$= \lambda(u, \mathscr{A}v) = \lambda(\mathscr{A}u, v) = \lambda(u,v)$$

for all $v \in H^1$, that is, solves the full equation (6.12).   □

In the limit case $\gamma = 0$, the modified eigenvalue problem therefore transfers again into the original eigenvalue equation (4.17) from which our discussion started.

6.2 Spaces of Functions with High-Order Mixed Derivatives

We attempt to prove that the solutions of the equation (6.11) possess, regardless of their origin, high-order mixed derivatives and that it is possible to estimate the L_2-norms of these derivatives by the L_2-norm of the solutions themselves. Let

$$\Delta_i = \sum_{k=1}^{3} \frac{\partial^2}{\partial x_{i,k}^2} \tag{6.13}$$

denote the Laplacian that acts on the spatial coordinates $x_{i,1}$, $x_{i,2}$, and $x_{i,3}$ of the electron i and let the differential operator $\mathscr{L}$ of order $2|I|$ be the product

$$\mathscr{L} = (-1)^{|I|} \prod_{i \in I} \Delta_i \tag{6.14}$$

of the second-order operators $-\Delta_i$. The seminorms $|\cdot|_{I,0}$ and $|\cdot|_{I,1}$ on the space $\mathscr{D}$ of the infinitely differentiable functions with compact support are then defined by

$$|u|_{I,0}^2 = (u, \mathscr{L}u), \quad |u|_{I,1}^2 = -(u, \Delta \mathscr{L}u). \tag{6.15}$$

Correspondingly, we introduce, for $s = 0, 1$, the norms given by

$$\|u\|_{I,s}^2 = \|u\|_s^2 + |u|_{I,s}^2. \tag{6.16}$$

Let I^* be the set of all mappings $\alpha : I \to \{1,2,3\}$. The operator $\mathscr{L}$ and with that the given seminorms can then be written in terms of the products

$$L_\alpha = \prod_{i \in I} \frac{\partial}{\partial x_{i,\alpha(i)}}, \quad \alpha \in I^*, \tag{6.17}$$

of first-order differential operators, more precisely as the sum

$$\mathscr{L} = (-1)^{|I|} \sum_{\alpha \in I^*} L_\alpha^2. \tag{6.18}$$

Correspondingly, since all partial derivatives of a function in $\mathscr{D}$ commute,

$$|u|_{I,0}^2 = \sum_{\alpha \in I^*} \|L_\alpha u\|_0^2, \quad |u|_{I,1}^2 = \sum_{\alpha \in I^*} |L_\alpha u|_1^2. \tag{6.19}$$

The completions of $\mathscr{D}_I$ under the norms given by (6.16) are the spaces X_I^s. They consist of functions that possess, for big $|I|$, very high order weak partial derivatives. We will show in that the solutions of the equation (6.11) are contained in X_I^1.

The structure of the proof of our regularity theorems is in the end very simple. Expressed naively, we transform the strong form

$$\tilde{H}u := Hu + \gamma \sum_{i=1}^N \theta_i \left\{ \frac{x_i}{|x_i|} \cdot \nabla_i u + \frac{1}{|x_i|} u \right\} = \lambda u, \tag{6.20}$$

of the second-order equation (6.12) into the high-order equation

$$(\varepsilon I + \mathscr{L}) \tilde{H} u = \lambda (\varepsilon I + \mathscr{L}) u \tag{6.21}$$

with correspondingly smooth solutions. As the operator $\varepsilon I + \mathscr{L}$ is invertible for $\varepsilon > 0$, both equations are equivalent and our regularity theorem is proved. Of course, this does not work in this simple way, one reason being all the singularities of the coefficient functions of the operator $\tilde{H}$. However, we can switch to the weak form

$$a(u, \varepsilon v + \mathscr{L}v) + \gamma s(u, \varepsilon v + \mathscr{L}v) = \lambda(u, \varepsilon v + \mathscr{L}v), \quad v \in \mathscr{D}_I, \tag{6.22}$$

of this equation, that is formally obtained from (6.21) if one multiplies both sides of the equation with a test function $v \in \mathscr{D}_I$, integrates, and then transforms the resulting integrals integrating by parts, or simply by replacing the test functions v in (6.12) by test functions $\varepsilon v + \mathscr{L}v$. The solutions of equation (6.12) obviously satisfy the equation (6.22). The idea is to interpret this equation as an equation on X_I^1 and to show that its solutions are conversely solutions of the original equation (6.12). Before we can realize this idea, we have, however, to show that the bilinear form

$$\tilde{a}(u,v) = a(u, \varepsilon v + \mathscr{L}v) + \gamma s(u, \varepsilon v + \mathscr{L}v) \tag{6.23}$$

on $\mathcal{D}_I \times \mathcal{D}_I$ can be extended to a bounded bilinear form on $X_I^1 \times X_I^1$. This is trivial for its leading part. The problem is to estimate its singular low-order terms correspondingly. The next two sections exclusively deal with this task.

6.3 Estimates for the Low-Order Terms, Part 1

As stated, the key to our regularity theory is estimates for the low-order terms in the bilinear form (6.23), that is, for the terms involving the interaction potentials

$$V_{ne}(x) = -\sum_{i=1}^{N} \sum_{v=1}^{K} \frac{Z_v}{|x_i - a_v|}, \quad V_{ee}(x) = \frac{1}{2} \sum_{\substack{i,j=1 \\ i \neq j}}^{N} \frac{1}{|x_i - x_j|} \tag{6.24}$$

between the nuclei and the electrons and between the electron among each other, and estimates for the part arising from the bilinear form (6.9). This bilinear form consists, like the nucleus-electron interaction potential, of a sum of one-electron terms. The terms involving only one single electron represent the simple part. The corresponding estimates are in the end based on the Hardy inequality from Lemma 4.1. They do not rely on symmetry properties of the wave functions. The situation is different for the terms of which the electron-electron interaction potential is composed. These estimates are therefore treated in a separate section.

The first of the estimates we need to study the regularity properties, namely the estimate (4.11) from Theorem 4.1, has already been stated in Chap. 4 and formed the basis of the variational formulation of the eigenvalue problem. The aim of the present section is to complement this estimate by estimates for the expressions

$$(V_{ne}u, \mathcal{L}v), \quad s(u, \mathcal{L}v), \quad s(u, v). \tag{6.25}$$

in the bilinear form (6.23) respectively in (6.11). The crucial observation is that most of the partial derivatives of which the differential operator $\mathcal{L}$ is composed commute with the single parts of the interaction potentials (6.24) and can be shifted from one to the other side in the single parts of the bilinear form (6.9), up to those few that act on a component of the position vectors of the electrons under consideration.

Theorem 6.3. *For all infinitely differentiable functions u and v in the space $\mathcal{D}$,*

$$(V_{ne}u, \mathcal{L}v) \leq 2N^{1/2}Z|u|_{I,0}|v|_{I,1}. \tag{6.26}$$

Proof. We first consider a single electron i and have then to distinguish the cases $i \notin I$ and $i \in I$. The first case is the easier one. We start from the representation (6.18) of $\mathcal{L}$. Since the partial derivatives of which the L_α are composed in this case do not act on the components of x_i, Fubini's theorem and integration by parts yield

$$\int \frac{1}{|x_i - a_v|} \, u \mathscr{L} v \, dx = (-1)^{|I|} \sum_{\alpha \in I^*} \int \frac{1}{|x_i - a_v|} \left(\int u L_\alpha^2 v \, d\tilde{x} \right) dx_i$$

$$= \sum_{\alpha \in I^*} \int \left(\int \frac{1}{|x_i - a_v|} L_\alpha u L_\alpha v \, dx_i \right) d\tilde{x},$$

where we have split x into x_i and $\tilde{x}$. By the Cauchy-Schwarz and the Hardy inequalities, the inner integrals on the right hand side can be estimated by the expressions

$$\left(\int |L_\alpha u|^2 \, dx_i \right)^{1/2} \left(4 \sum_{\ell=1}^{3} \int |\frac{\partial}{\partial x_{i,\ell}} L_\alpha v|^2 \, dx_i \right)^{1/2}.$$

With help of the Cauchy-Schwarz inequality, now first applied to the resulting outer integrals and then to the sum over the single $\alpha \in I^*$, the estimate

$$\int \frac{1}{|x_i - a_v|} \, u \mathscr{L} v \, dx$$

$$\leq 2 \left(\sum_{\alpha \in I^*} \int |L_\alpha u|^2 \, dx \right)^{1/2} \left(\sum_{\alpha \in I^*} \sum_{\ell=1}^{3} \int |\frac{\partial}{\partial x_{i,\ell}} L_\alpha v|^2 \, dx \right)^{1/2}$$

follows. In more compact notion, this estimate reads

$$\int \frac{1}{|x_i - a_v|} \, u \mathscr{L} v \, dx \leq 2 \, |u|_{I,0} |\nabla_i v|_{I,0}. \tag{6.27}$$

It transfers without change to the case of indices $i \in I$, but the proof is somewhat more complicated then. In this case, we decompose the operator $\mathscr{L}$ into the sum

$$\mathscr{L} = (-1)^{|I|} \sum_{\alpha \in I^*} L_\alpha^2 = (-1)^{|I|} \sum_{\beta \in I_i^*} L_\beta \Delta_i L_\beta, \quad L_\beta = \prod_{j \in I_i} \frac{\partial}{\partial x_{j,\beta(j)}},$$

where $I_i = I \setminus \{i\}$ and I_i^* denotes the set of the mappings β that assign one of the components 1, 2, or 3 to the electron indices j in I_i. Since the L_β do not act upon the components of x_i, integration by parts and Fubini's theorem lead as above to

$$\int \frac{1}{|x_i - a_v|} \, u \mathscr{L} v \, dx = (-1)^{|I|} \sum_{\beta \in I_i^*} \int \frac{1}{|x_i - a_v|} \left(\int u L_\beta \Delta_i L_\beta v \, d\tilde{x} \right) dx_i$$

$$= - \sum_{\beta \in I_i^*} \int \left(\int \frac{1}{|x_i - a_v|} L_\beta u \Delta_i L_\beta v \, dx_i \right) d\tilde{x}.$$

By the Cauchy-Schwarz and the Hardy inequality, the inner integrals on the right hand side can, up to the factor 2, be estimated by the expressions

$$\left(\int |\nabla_i L_\beta u|^2 \, dx_i \right)^{1/2} \left(\int |\Delta_i L_\beta v|^2 \, dx_i \right)^{1/2}.$$

These expressions can be rewritten as

$$\left(\sum_{k=1}^{3} \int \left| \frac{\partial L_\beta u}{\partial x_{i,k}} \right|^2 dx_i \right)^{1/2} \left(\sum_{k=1}^{3} \sum_{\ell=1}^{3} \int \left| \frac{\partial}{\partial x_{i,\ell}} \frac{\partial L_\beta v}{\partial x_{i,k}} \right|^2 dx_i \right)^{1/2},$$

where we have applied the relation

$$\sum_{k=1}^{3} \sum_{\ell=1}^{3} \int \frac{\partial^2 w}{\partial x_{i,k}^2} \frac{\partial^2 w}{\partial x_{i,\ell}^2} dx_i = \sum_{k=1}^{3} \sum_{\ell=1}^{3} \int \left| \frac{\partial^2 w}{\partial x_{i,\ell} \partial x_{i,k}} \right|^2 dx_i$$

to the functions $w = L_\beta v$. This relation is proved by integrating by parts. Since the set of the differential operators L_α, $\alpha \in I^*$, coincides with the set of the operators

$$\frac{\partial}{\partial x_{i,k}} L_\beta, \quad k = 1,2,3, \ \beta \in I_i^*,$$

summation over all β, the Cauchy-Schwarz inequality (applied twice, to the outer integrals and then to the sum over the β), and Fubini's theorem lead again to (6.27).

Summation over the single contributions in the potential finally yields

$$(V_{ne}u, \mathscr{L}v) \leq 2Z|u|_{I,0} \sum_{i=1}^{N} |\nabla_i v|_{I,0},$$

from which the proposition follows with the elementary estimate

$$\sum_{i=1}^{N} |\nabla_i v|_{I,0} \leq N^{1/2} \left(\sum_{i=1}^{N} |\nabla_i v|_{I,0}^2 \right)^{1/2} = N^{1/2} |v|_{I,1},$$

that is responsible for the factor $N^{1/2}$. □

The proof of the estimates for the expression $s(u, \mathscr{L}v)$ resembles that of Theorem 6.3. It is prepared by the following lemma for functions of three real variables.

Lemma 6.3. *For all infinitely differentiable functions $u, v : \mathbb{R}^3 \to \mathbb{R}$ that vanish outside a bounded subset of their domain,*

$$\int \left\{ \frac{x}{|x|} \cdot \nabla u + \frac{1}{|x|} u \right\} v \, dx \leq 3 \left(\int |u|^2 \, dx \right)^{1/2} \left(\int |\nabla v|^2 \, dx \right)^{1/2}. \tag{6.28}$$

Proof. The difficulty is that the derivatives have to be shifted to v. We first assume that u vanishes on a neighborhood of the origin. Integration by parts then yields

$$\int \left\{ \frac{x}{|x|} \cdot \nabla u + \frac{1}{|x|} u \right\} v \, dx = -\int u \frac{x}{|x|} \cdot \nabla v \, dx - \int \frac{1}{|x|} u v \, dx.$$

This relation remains true for the general case, as one can show by an argument as in the proof of Lemma 4.1, that is, by multiplying u with a sequence of cut-off functions and applying the dominated convergence theorem. The proposition then follows again from the Cauchy-Schwarz inequality and the Hardy inequality. □

Theorem 6.4. *For all infinitely differentiable functions u and v in the space $\mathscr{D}$,*

$$s(u, \mathscr{L}v) \le 3 \, |u|_{1,0} \, |v|_{1,1}. \qquad (6.29)$$

Proof. We consider again a single electron i and have, as in the proof of Theorem 6.3, to distinguish the cases $i \in I$ and $i \notin I$. For indices $i \in I$, one obtains

$$\int \left\{ \frac{x_i}{|x_i|} \cdot \nabla_i u + \frac{1}{|x_i|} u \right\} \mathscr{L}v \, dx$$
$$= \sum_{\beta \in I_i^*} \iint \left\{ \frac{x_i}{|x_i|} \cdot \nabla_i L_\beta u + \frac{1}{|x_i|} L_\beta u \right\} \Delta_i L_\beta v \, dx_i \, d\tilde{x}.$$

With help of the Cauchy-Schwarz and the Hardy inequality the inner integrals on the right hand side can, up to the factor 3, be estimated by the expressions

$$\left(\int |\nabla_i L_\beta u|^2 \, dx_i \right)^{1/2} \left(\int |\Delta_i L_\beta v|^2 \, dx_i \right)^{1/2}.$$

Rewriting these expressions as in the proof of Theorem 6.3, from this the estimate

$$\int \left\{ \frac{x_i}{|x_i|} \cdot \nabla_i u + \frac{1}{|x_i|} u \right\} \mathscr{L}v \, dx \le 3 \, |u|_{1,0} \, |\nabla_i v|_{1,0}$$

follows. This estimate also holds if $i \notin I$, as is shown starting directly from the representation of $\mathscr{L}$ as the sum of the differential operators L_α^2, that is, from

$$\int \left\{ \frac{x_i}{|x_i|} \cdot \nabla_i u + \frac{1}{|x_i|} u \right\} \mathscr{L}v \, dx$$
$$= \sum_{\alpha \in I^*} \iint \left\{ \frac{x_i}{|x_i|} \cdot \nabla_i L_\alpha u + \frac{1}{|x_i|} L_\alpha u \right\} L_\alpha v \, dx_i \, d\tilde{x}.$$

The inner integrals are now, with Lemma 6.3, up to the factor 3 estimated as

$$\left(\int |L_\alpha u|^2 \, dx_i \right)^{1/2} \left(\int |\nabla_i L_\alpha v|^2 \, dx_i \right)^{1/2}.$$

From that then again the estimate above follows. Summation over the i, the Cauchy-Schwarz inequality, and the fact that the θ_i^2 sum up to 1 complete the proof. □

The group of estimates for the one-electron parts in the bilinear form (6.23) is completed by the following estimate for the expression $s(u,v)$ itself:

Theorem 6.5. *For all infinitely differentiable functions u and v in the space $\mathcal{D}$,*

$$s(u,v) \leq 3\,\|u\|_0\,|v|_1. \tag{6.30}$$

Proof. With help of Lemma 6.3, the single parts can again be estimated as

$$\int \left\{ \frac{x_i}{|x_i|} \cdot \nabla_i u + \frac{1}{|x_i|}\, u \right\} v\, dx \leq 3\,\|u\|_0\,\|\nabla_i v\|_0.$$

The proposition follows from that in the way already employed. □

6.4 Estimates for the Low-Order Terms, Part 2

The part in the bilinear form resulting from the electron-electron interaction potential is estimated basically in the same way as the terms considered in the previous section. The central observation is again that most of the derivatives of which the differential operators L_α are composed commute with the single parts of the potential. However, there is one important difference. In the cases already studied only one derivative remained, in contrast to the two derivatives we have to face here. One of these derivatives has to be shifted to the other side. This causes an additional problem since the partial derivatives of the interaction potential entering into the estimates are not locally square integrable in three space dimensions. Therefore the Pauli principle has to be brought into play. A wave function that is compatible with the Pauli principle vanishes where two electrons with the same spin meet, a fact which counterbalances the singular behavior of the derivatives of the interaction potential and enables us to estimate the terms under consideration.

To master the most singular terms, the Hardy estimate from Lemma 4.1 has to be complemented by a second, closely related estimate for functions of three variables.

Lemma 6.4. *For all infinitely differentiable functions v in the variable $x \in \mathbb{R}^3$ that have a compact support and that vanish at the origin,*

$$\int \frac{1}{|x|^4} v^2\, dx \leq 4 \int \frac{1}{|x|^2} |\nabla v|^2\, dx. \tag{6.31}$$

Proof. The estimate is proved in the same way as the Hardy inequality (4.8). Setting temporarily $d(x) = |x|$, it starts from the relation

$$\frac{1}{d^4} = -\frac{1}{3} \nabla\!\left(\frac{1}{d^3}\right) \cdot \nabla d,$$

with the help of which (6.31) is proved for functions v that vanish on a neighborhood of the origin. To transfer this estimate to functions v that vanish only at the origin itself, one has to utilize that in this case there exists a constant K with

$$|v(x)| \leq K|x|$$

and can then complete the proof in the same way as that of (4.8) with help of the dominated convergence theorem, multiplying v with cut-off functions. □

It should be noted that the estimate (6.31) does not hold for functions not vanishing at the origin since the function $x \to 1/|x|^4$ is not locally integrable in three space dimensions, which is the source of our problems.

The single parts of which the electron-electron interaction potential is composed involve only two electrons so that the estimates that we have to prove are essentially two-electron estimates. To simplify the notation, we restrict ourselves for a while to the two-electron case and denote the three-dimensional coordinate vectors of these electrons by x and y. Correspondingly, the real numbers x_1, x_2, and x_3 and y_1, y_2, and y_3 are the components of these vectors. For abbreviation, let

$$\phi(x,y) = \frac{1}{|x-y|}. \tag{6.32}$$

In this notation, our task is essentially to estimate the integrals like

$$\int \phi u \sum_{k,\ell=1}^{3} \frac{\partial^4 v}{\partial x_k^2 \partial y_\ell^2} \, d(x,y) \tag{6.33}$$

for infinitely differentiable functions u and v that have a compact support and that are antisymmetric under the exchange of x and y.

The first step is to combine the inequality (6.31) and the Hardy inequality (4.8) to the estimate for antisymmetric functions on which our argumentation is founded.

Lemma 6.5. *For all infinitely differentiable functions u in the variables $x,y \in \mathbb{R}^3$ that have a compact support and are antisymmetric under the exchange of x and y,*

$$\int \frac{1}{|x-y|^4} u^2 \, d(x,y) \leq 16 \sum_{k,\ell=1}^{3} \int \left(\frac{\partial^2 u}{\partial x_k \partial y_\ell} \right)^2 d(x,y). \tag{6.34}$$

Proof. Since such functions vanish where $y = x$, Lemma 6.4 yields

$$\int \left(\int \frac{1}{|x-y|^4} u^2 \, dy \right) dx \leq \int \left(4 \sum_{\ell} \int \frac{1}{|x-y|^2} \left(\frac{\partial u}{\partial y_\ell} \right)^2 dy \right) dx.$$

By the Hardy inequality from Lemma 4.1,

$$\int \left(\int \frac{1}{|x-y|^2} \left(\frac{\partial u}{\partial y_\ell} \right)^2 dx \right) dy \leq \int \left(4 \sum_k \int \left(\frac{\partial^2 u}{\partial x_k \partial y_\ell} \right)^2 dx \right) dy.$$

The proposition follows with Fubini's theorem. □

The counterparts to this estimate are the following variants

$$\int \frac{1}{|x-y|^2} v^2 \, d(x,y) \leq 4 \sum_{k=1}^{3} \int \left(\frac{\partial v}{\partial x_k} \right)^2 d(x,y), \tag{6.35}$$

$$\int \frac{1}{|x-y|^2} v^2 \, d(x,y) \leq 4 \sum_{\ell=1}^{3} \int \left(\frac{\partial v}{\partial y_\ell} \right)^2 d(x,y) \tag{6.36}$$

of the Hardy inequality (4.8) that, in contrast to (6.34), do not rely on the antisymmetry of the considered function. They are proved in the same way as (6.34). The argumentation in this section centers in the estimates (6.34), (6.35), and (6.36).

Now we can begin to estimate the integrals (6.33). In the first step we shift one of the partial derivatives from the function v to the function u.

Lemma 6.6. *Let u and v be infinitely differentiable functions in the variables x and y in $\mathbb{R}^3$ that have a compact support. Then, for all indices k and ℓ,*

$$\int \phi u \frac{\partial^4 v}{\partial x_k^2 \partial y_\ell^2} \, d(x,y) = -\int \frac{\partial}{\partial x_k} (\phi u) \frac{\partial^3 v}{\partial x_k \partial y_\ell^2} \, d(x,y). \tag{6.37}$$

Proof. The problem is the singularity of ϕ that does not allow to integrate by parts directly. Let $\varphi(r)$ thus be a continuously differentiable function of the real variable $r \geq 0$ that coincides with the function $1/r$ for $r \geq 1$ and is constant for $r \leq 1/2$. Let

$$\phi_n(x,y) = n \, \varphi(n \, |x-y|), \quad n \in \mathbb{N}.$$

The ϕ_n are then itself continuously differentiable and coincide with the original function ϕ for all x and y of distance $|x-y| \geq 1/n$. Integration by parts leads to

$$\int \phi_n u \frac{\partial^4 v}{\partial x_k^2 \partial y_\ell^2} \, d(x,y) = -\int \frac{\partial}{\partial x_k} (\phi_n u) \frac{\partial^3 v}{\partial x_k \partial y_\ell^2} \, d(x,y).$$

The integral on the right hand side of this equation splits, because of

$$\frac{\partial}{\partial x_k} (\phi_n u) = \frac{\partial \phi_n}{\partial x_k} u + \phi_n \frac{\partial u}{\partial x_k},$$

into two parts. We claim that there is a constant M, independent of n, such that

$$\left| \frac{\partial}{\partial x_k} (\phi_n u) \right| \leq \frac{M}{|x-y|^2}.$$

This is because, for the function ϕ_n itself and its first-order derivatives, the estimates

$$|\phi_n| \leq \frac{c}{|x-y|}, \quad \left| \frac{\partial \phi_n}{\partial x_k} \right| \leq \frac{c}{|x-y|^2},$$

hold, where c is independent of n. As u vanishes outside a bounded set, the integrands are thus uniformly bounded by an integrable function. Since the ϕ_n and their first-order partial derivatives converge to ϕ and its respective derivatives outside the diagonal $x = y$, a set of measure zero, the dominated convergence theorem yields

$$\lim_{n \to \infty} \int \frac{\partial}{\partial x_k} (\phi_n u) \frac{\partial^3 v}{\partial x_k \partial y_\ell^2} \, d(x, y) = \int \frac{\partial}{\partial x_k} (\phi u) \frac{\partial^3 v}{\partial x_k \partial y_\ell^2} \, d(x, y).$$

For the other side of the equation, one can argue correspondingly and obtains

$$\lim_{n \to \infty} \int \phi_n u \frac{\partial^4 v}{\partial x_k^2 \partial y_\ell^2} \, d(x, y) = \int \phi u \frac{\partial^4 v}{\partial x_k^2 \partial y_\ell^2} \, d(x, y),$$

which then completes the proof of (6.37). $\square$

The next estimate is the place where the antisymmetry crucially enters. It depends on the fact that the corresponding functions u vanish on the diagonal $x = y$.

Lemma 6.7. *Let u and v be infinitely differentiable functions in the variables x, y in $\mathbb{R}^3$ that have a compact support and let the function u be antisymmetric with respect to the exchange of x and y. Then the estimate*

$$\sum_{k,\ell=1}^{3} \int \phi u \frac{\partial^4 v}{\partial x_k^2 \partial y_\ell^2} \, d(x, y) \tag{6.38}$$

$$\leq C \left\{ \sum_{k,\ell=1}^{3} \left\| \frac{\partial^2 u}{\partial x_k \partial y_\ell} \right\|_0^2 \right\}^{1/2} \left\{ \sum_{k,\ell=1}^{3} \left| \frac{\partial^2 v}{\partial x_k \partial y_\ell} \right|_1^2 \right\}^{1/2},$$

holds, where the constant C is specified in the proof.

Proof. We first rewrite the expression to be estimated with help of (6.37) and obtain

$$-\sum_{k,\ell=1}^{3} \int \frac{1}{|x-y|} \frac{\partial u}{\partial x_k} \frac{\partial^3 v}{\partial x_k \partial y_\ell^2} \, d(x, y) + \sum_{k,\ell=1}^{3} \int \frac{1}{|x-y|^2} \frac{x_k - y_k}{|x-y|} u \frac{\partial^3 v}{\partial x_k \partial y_\ell^2} \, d(x, y).$$

The first double sum is estimated by the expression

$$\left(3 \sum_{k=1}^{3} \int \frac{1}{|x-y|^2} \left(\frac{\partial u}{\partial x_k} \right)^2 d(x, y) \right)^{1/2} \left(\sum_{k,\ell=1}^{3} \int \left(\frac{\partial^3 v}{\partial x_k \partial y_\ell^2} \right)^2 d(x, y) \right)^{1/2}.$$

As u vanishes on the diagonal $x = y$, there is a constant K with

$$|u(x, y)| \leq K |x - y|.$$

The second double sum is thus bounded by the therefore finite expression

$$\left(3 \int \frac{1}{|x-y|^4} u^2 \, d(x, y) \right)^{1/2} \left(\sum_{k,\ell=1}^{3} \int \left(\frac{\partial^3 v}{\partial x_k \partial y_\ell^2} \right)^2 d(x, y) \right)^{1/2}.$$

The estimates (6.36), applied to the partial derivatives of u, and (6.34) show that the estimate (6.38) holds with $C = 6\sqrt{3}$. Since the role of x and y can be exchanged, the constant can be improved to $C = 3\sqrt{6}$, combining the two resulting estimates. □

Correspondingly one proves the estimate

$$\sum_{k=1}^{3} \int \phi u \frac{\partial^2 v}{\partial x_k^2} \, d(x,y) \leq 2 \left\{ \sum_{k=1}^{3} \left\| \frac{\partial u}{\partial x_k} \right\|_0^2 \right\}^{1/2} \left\{ \sum_{k=1}^{3} \left| \frac{\partial v}{\partial x_k} \right|_1^2 \right\}^{1/2} \tag{6.39}$$

applying (6.35) to u, and finally, with help of (6.35) and (6.36), the estimate

$$\int \phi u v \, d(x,y) \leq \sqrt{2} \, \|u\|_0 |v|_1 \tag{6.40}$$

for all infinitely differentiable functions u and v that have a compact support, in these cases regardless their antisymmetry with respect to the exchange of x and y.

We can now return to the full set of the electron coordinate vectors $x_1, x_2, \ldots, x_N$ in $\mathbb{R}^3$ and the old notation and merge the building blocks (6.38) to (6.40) into the last missing estimate for the interaction potentials.

Theorem 6.6. *For all infinitely differentiable functions $u \in \mathscr{D}_I$ and $v \in \mathscr{D}$,*

$$(V_{ee}u, \mathscr{L}v) \leq CN^{3/2} |u|_{I,0} |v|_{I,1}, \tag{6.41}$$

where the constant $C \leq 3\sqrt{3}$ is independent of the number N of electrons.

Proof. We first turn our attention to the interaction potential

$$\phi_{ij}(x) = \frac{1}{|x_i - x_j|}$$

of two electrons $i \neq j$ and estimate the expression

$$\int \phi_{ij} u \, \mathscr{L}v \, dx = (-1)^{|I|} \sum_{\alpha \in I^*} \int \phi_{ij} u \, L_\alpha^2 v \, dx.$$

The strategy is the same as in the previous section. We split the operators L_α into the product of operators L_β that do not act upon the components of x_i and x_j and a remaining part. Here we have to distinguish three cases, namely that both indices i and j belong to the index set I, that only one of these indices belongs to I, and that none of these indices is contained in I.

The first case is the most critical one because of the singularities of the derivatives of the interaction potential and the dependence on the antisymmetry. It is therefore considered first. Let $I_{ij} = I \setminus \{i, j\} \neq \emptyset$ and let I_{ij}^* again denote the set of the mappings β that assign one of the components 1, 2, or 3 to an electron index in I_{ij}. The set of the differential operators L_α, $\alpha \in I^*$, coincides then with the set of the operators

$$\frac{\partial}{\partial x_{i,k}} \frac{\partial}{\partial x_{i,\ell}} L_\beta, \quad k,\ell = 1,2,3, \ \beta \in I_{ij}^*,$$

and the integral to be estimated can, as in the previous section, be written as sum

$$(-1)^{|I|} \sum_{\alpha \in I^*} \int \phi_{ij} u L_\alpha^2 v \, dx = \sum_{\beta \in I_{ij}^*} \int \left(\sum_{k,l=1}^{3} \iint \phi_{ij} L_\beta u \, \frac{\partial^4 L_\beta v}{\partial x_{i,k}^2 \partial x_{j,\ell}^2} \, dx_i dx_j \right) d\tilde{x},$$

where x is split into x_i, x_j, and the remaining components $\tilde{x}$. Like u itself, its partial derivatives $L_\beta u$, $\beta \in I_{ij}^*$, are antisymmetric under the exchange of x_i and x_j. This is due to the fact that the operators L_β do not act upon the components of x_i and x_j and can be seen as follows. Let w be an arbitrary function that changes its sign under the permutation P that exchanges x_i for x_j and let $e \neq 0$ be a vector that is invariant under P. Let $\tilde{w}(x) = w(Px)$. Since $e = Pe$ and $\tilde{w}(x) = -w(x)$, then

$$(\nabla w)(Px) \cdot e = P^T (\nabla w)(Px) \cdot e = (\nabla \tilde{w})(x) \cdot e = - (\nabla w)(x) \cdot e,$$

so that the directional derivative of w in direction e inherits the antisymmetry of w. The proposition follows from that by induction on the order of L_β. The inner integrals on the right hand side of the equation above can therefore be estimated with the help of (6.38). In the same fashion as in the previous section, finally the estimate

$$(-1)^{|I|} \sum_{\alpha \in I^*} \int \phi_{ij} u L_\alpha^2 v \, dx \leq C |u|_{I,0} \left\{ |\nabla_i v|_{I,0}^2 + |\nabla_j v|_{I,0}^2 \right\}^{1/2} \qquad (6.42)$$

follows, where $C \leq 3\sqrt{6}$ is the same constant as in (6.38). The case that I_{ij} is empty, that is, I consists only of the indices i and j, is treated in the same way.

In the case that $i \in I$, but $j \notin I$, we set $I_i = I \setminus \{i\}$ and denote by I_i^* again the set of the mappings β from I_i to the set of the indices 1, 2, and 3. The set of the differential operators L_α, $\alpha \in I^*$, then coincides with the set of the operators

$$\frac{\partial}{\partial x_{i,k}} L_\beta, \quad k = 1,2,3, \ \beta \in I_i^*,$$

and the integral to be estimated splits into the sum

$$(-1)^{|I|} \sum_{\alpha \in I^*} \int \phi_{ij} u L_\alpha^2 v \, dx = - \sum_{\beta \in I_i^*} \int \left(\sum_{k=1}^{3} \iint \phi_{ij} L_\beta u \, \frac{\partial^2 L_\beta v}{\partial x_{i,k}^2} \, dx_i dx_j \right) d\tilde{x}.$$

The inner sum on the right hand side can be estimated with help of (6.39), which then finally again results in the estimate (6.42), where $C \leq 2$ is now the constant from (6.39). The same estimate holds, of course, for the case that $i \notin I$ and $j \in I$.

If neither i nor j are contained in I, one simply starts from

$$(-1)^{|I|} \sum_{\alpha \in I^*} \int \phi_{ij} u L_\alpha^2 v \, dx = \sum_{\alpha \in I^*} \int \left(\iint \phi_{ij} L_\alpha u L_\alpha v \, dx_i dx_j \right) d\tilde{x},$$

from which one obtains, with the help of (6.40), again the estimate (6.42), now with a constant $C \leq \sqrt{2}$. Independent of whether two, one, or none of the indices i and j is contained in I, the estimate (6.42) holds with a constant $C \leq 3\sqrt{6}$.

The proposition finally follows from the elementary estimate

$$\frac{1}{2} \sum_{i,j} (\eta_i^2 + \eta_j^2)^{1/2} \leq \frac{1}{\sqrt{2}} N^{3/2} \left(\sum_i \eta_i^2 \right)^{1/2}, \tag{6.43}$$

summing over all particle pairs. □

Again, the dependence of the bound on the problem parameters, here the number N of electrons, enters only in the very last step, through the estimate (6.43).

6.5 The Regularity of the Weighted Eigenfunctions

We are now in the position to prove that the solutions $u \in H_I^1$ of the modified eigenvalue equation (6.11) are located in the space X_I^1 from Sect. 6.2, the completion of the space $\mathscr{D}_I$ of the infinitely differentiable functions (4.3) with compact support that are antisymmetric under the exchange of arguments x_i and x_j in $\mathbb{R}^3$ for all indices $i \neq j$ in the given subset I of the set of indices $1, \ldots, N$ under a norm measuring high-order mixed derivatives. The key to our results is the estimates for the low-order terms, those discussed in the preceding two sections, that can be summarized as follows. For all functions u in $\mathscr{D}_I$ and v in $\mathscr{D}$, first the estimates

$$(Vu, \mathscr{L}v) \leq C\theta(N,Z) |u|_{I,0} |v|_{I,1}, \quad s(u, \mathscr{L}v) \leq 3 |u|_{I,0} |v|_{I,1} \tag{6.44}$$

in terms of the seminorms (6.15) hold, where the first one for the term with the interaction potential (4.9) represents a combination of the estimates (6.26) from Theorem 6.3 and (6.41) from Theorem 6.6, and the second one is the estimate (6.29) from Theorem 6.4. The constant C is independent of the number N of electrons, of the considered index set I, of the number, the position, and the charge of the nuclei, and particularly of their total charge Z. The proofs yielded the upper bound $C = 2 + 3\sqrt{3}$ for C. The quantity $\theta(N,Z)$ has been defined in (4.10) and covers the growth of the bound in N and Z. The antisymmetry of the functions u with respect to the exchange of the corresponding electron coordinates substantially enters into the proof of the first estimate, since without this property it is not possible to get a handle on the electron-electron interaction terms. The estimates (6.44) potentially involving very high-order derivatives are complemented by the estimates

$$(Vu, v) \leq 3\theta(N,Z) \|u\|_0 |v|_1, \quad s(u,v) \leq 3 \|u\|_0 |v|_1 \tag{6.45}$$

from Theorem 4.1 and Theorem 6.5 for functions u and v in $\mathscr{D}$, that generally hold and do not rely on the given antisymmetry properties. The estimates show that the bilinear forms $(Vu, \mathscr{L}v)$ and $s(u, \mathscr{L}v)$ can be uniquely extended from $\mathscr{D}_I \times \mathscr{D}_I$ to bounded bilinear forms on $X_I^0 \times X_I^1$, and that particularly the bilinear form

$$\tilde{a}(u, v) = a(u, \varepsilon v + \mathscr{L}v) + \gamma s(u, \varepsilon v + \mathscr{L}v) \tag{6.46}$$

from Sect. 6.2 can be uniquely extended from $\mathscr{D}_I$ to a bounded bilinear form on X_I^1. For the ease of presentation, we will keep the notation $(Vu, \mathscr{L}v)$ and $s(u, \mathscr{L}v)$ for arguments $u \in X_I^0$ and $v \in X_I^1$ and mean the extended forms then, where, of course, some care has to be taken to avoid misinterpretations and fallacies.

The second ingredient of the proof of the regularity theorems is Fourier analysis. Recall from Chap. 2 the definition of the space $\mathscr{S}$ of the rapidly decreasing functions. As with $\mathscr{D}_I$, let $\mathscr{S}_I$ denotes the space of the rapidly decreasing functions of corresponding antisymmetry. The seminorms (6.15) of a rapidly decreasing function read in terms of its Fourier transform

$$|u|_{I,s}^2 = \int \left(\sum_{i=1}^N |\omega_i|^2 \right)^s \left(\prod_{i \in I} |\omega_i|^2 \right) |\hat{u}(\omega)|^2 \, d\omega. \tag{6.47}$$

Correspondingly, the H^1-seminorm $|u|_1$ and the L_2-norm $\|u\|_0 = |u|_0$ are given by

$$|u|_s^2 = \int \left(\sum_{i=1}^N |\omega_i|^2 \right)^s |\hat{u}(\omega)|^2 \, d\omega. \tag{6.48}$$

We call a rapidly decreasing function a rapidly decreasing high-frequency function if its Fourier transform vanishes on a ball of radius Ω, to be fixed later, around the origin of the frequency space. The closures of the corresponding space

$$\mathscr{S}_{I,H} = \{ v \in \mathscr{S}_I \mid \hat{v}(\omega) = 0 \text{ for } |\omega| \le \Omega \} \tag{6.49}$$

of rapidly decreasing functions with the given symmetry properties in H_I^1 and X_I^1, respectively, are the Hilbert spaces $H_{I,H}^1$ and $X_{I,H}^1$. The closures of the space

$$\mathscr{S}_{I,L} = \{ v \in \mathscr{S}_I \mid \hat{v}(\omega) = 0 \text{ for } |\omega| \ge \Omega \} \tag{6.50}$$

in H_I^1 and X_I^1 are the spaces $H_{I,L}^1$ and $X_{I,L}^1$, respectively, of low-frequency functions. The low-frequency and the high-frequency functions decompose the spaces

$$H_I^1 = H_{I,L}^1 \oplus H_{I,H}^1, \quad X_I^1 = X_{I,L}^1 \oplus X_{I,H}^1 \tag{6.51}$$

into orthogonal parts. By the Fourier representation (6.47) and (6.48) of the norms,

$$|u_L|_{I,s} \le \Omega^s \left(\frac{\Omega}{\sqrt{|I|}} \right)^{|I|} \|u_L\|_0 \tag{6.52}$$

for the low-frequency functions $u_L \in \mathcal{S}_{I,L}$. The space $H_{I,L}^1$ and its subspace $X_{I,L}^1$ therefore coincide. The relation (6.52) transfers to all functions in these spaces. In fact, the functions in $H_{I,L}^1$ are infinitely differentiable and all their derivatives are square integrable. Fourier analysis also shows that

$$\|u_H\|_0 \leq \Omega^{-1} |u_H|_1, \quad |u_H|_{I,0} \leq \Omega^{-1} |u_H|_{I,1} \tag{6.53}$$

for all high-frequency functions in u_H in $H_{I,H}^1$ and $X_{I,H}^1$ respectively. On $H_{I,H}^1$, the seminorm $|\cdot|_1$ and the norm $\|\cdot\|_1$ thus are equivalent. For $u_L \in H_{I,L}^1$, conversely

$$|u_L|_1 \leq \Omega \|u_L\|_0, \quad |u_L|_{I,1} \leq \Omega |u_L|_{I,0}. \tag{6.54}$$

The central observation, on which the proof of the regularity theorems is based, is that the low-order terms in the bilinear form in the second-order equation (6.11), as well as in the high-order bilinear form (6.23), behave like small perturbations on the corresponding spaces of high-frequency functions. The reason is that the norms of such functions themselves and that of their derivatives as well can be estimated by the norms of derivatives of higher order. By (6.44) and (6.53),

$$(V u_H, \mathcal{L} v_H) \leq C \theta(N,Z) \Omega^{-1} |u_H|_{I,1} |v_H|_{I,1}, \tag{6.55}$$

$$s(u_H, \mathcal{L} v_H) \leq 3 \Omega^{-1} |u_H|_{I,1} |v_H|_{I,1} \tag{6.56}$$

for all $u_H, v_H \in \mathcal{S}_{I,H}$. Correspondingly, by (6.45) and (6.53), for these u_H and v_H

$$(V u_H, v_H) \leq 3 \theta(N,Z) \Omega^{-1} |u_H|_1 |v_H|_1, \tag{6.57}$$

$$s(u_H, v_H) \leq 3 \Omega^{-1} |u_H|_1 |v_H|_1. \tag{6.58}$$

This implies that the two bilinear forms become coercive on the corresponding spaces of high-frequency functions, provided that the bound Ω separating the low from the high frequencies is chosen large enough. If we assume $C \geq 3$ and choose

$$\Omega \geq 4 C \theta(N,Z) + 12\gamma, \tag{6.59}$$

for all high-frequency functions $u_H \in H_{I,H}^1$ the estimate

$$a(u_H, u_H) + \gamma s(u_H, u_H) \geq \frac{1}{4} |u_H|_1^2 \tag{6.60}$$

holds, and correspondingly, for the functions $u_H \in X_{I,H}^1$, the estimate

$$\tilde{a}(u_H, u_H) \geq \frac{1}{4} \left(\varepsilon |u_H|_1^2 + |u_H|_{I,1}^2 \right). \tag{6.61}$$

The claimed coercivity follows from that by the equivalence of the seminorm $|\cdot|_1$ and the norm $\|\cdot\|_1$ on the given spaces of high-frequency functions. We still combine the low-order terms in $\tilde{a}(u,v)$, respectively $a(u,v)$, in the bilinear forms

$$\widetilde{b}(\varphi, v) = (V\varphi, \varepsilon v + \mathscr{L}v) + \gamma s(\varphi, \varepsilon v + \mathscr{L}v), \qquad (6.62)$$

$$b(\varphi, \chi) = (V\varphi, \chi) + \gamma s(\varphi, \chi) \qquad (6.63)$$

on $X_I^0 \times X_I^1$ and $L_2 \times H^1$, respectively. They satisfy, for Ω as in (6.59), the estimates

$$\widetilde{b}(\varphi, v) \le \frac{1}{4}\Omega \left(\varepsilon \|\varphi\|_0^2 + |\varphi|_{I,0}^2 \right)^{1/2} \left(\varepsilon |v|_1^2 + |v|_{I,1}^2 \right)^{1/2}, \qquad (6.64)$$

$$b(\varphi, \chi) \le \frac{1}{4}\Omega \|\varphi\|_0 |\chi|_1 \qquad (6.65)$$

for functions φ, v, and χ in the corresponding spaces.

Due to the orthogonality properties of the low- and the high-frequency functions, the low- and the high-frequency part of a solution of the eigenvalue equation (6.11)

$$a(u, \chi) + \gamma s(u, \chi) = \lambda(u, \chi), \quad \chi \in H_I^1, \qquad (6.66)$$

interact only by the low-order part in the bilinear form on the left hand side. The aim is to control the high-frequency part and its mixed derivatives by the low-frequency part of the given solution. The first step to reach this goal is the following lemma that immediately results from the orthogonality of the low- and the high-frequency functions both with respect to the L_2- and the H^1-inner product.

Lemma 6.8. *Let $u = u_L + u_H$ be the decomposition of a solution $u \in H_I^1$ of the equation (6.11), (6.66) into its low-frequency and its high-frequency part. Then*

$$a(u_H, \chi_H) + \gamma s(u_H, \chi_H) - \lambda(u_H, \chi_H) = -b(u_L, \chi_H), \quad \chi_H \in H_{I,H}^1. \qquad (6.67)$$

We will keep the low-frequency part u_L fixed for a while and will consider (6.67) as an equation for the high-frequency part u_H. We will show that such equations are uniquely solvable for frequency bounds (6.59) and that the regularity of the right hand side transfers to the regularity of the solution.

Lemma 6.9. *For frequency bounds Ω as in (6.59), the equation*

$$a(u_H, \chi_H) + \gamma s(u_H, \chi_H) + \mu(u_H, \chi_H) = b(\varphi, \chi_H), \quad \chi_H \in H_{I,H}^1, \qquad (6.68)$$

possesses a unique solution $u_H \in H_{I,H}^1$ for all given functions $\varphi \in L_2$ and arbitrary nonnegative parameters μ. This solution satisfies the estimates

$$\|u_H\|_0 \le \|\varphi\|_0, \quad |u_H|_1 \le \Omega \|\varphi\|_0. \qquad (6.69)$$

Proof. As $\mu \ge 0$, the additional term does not alter the coercivity (6.60) of the bilinear form on the left hand side of the equation (6.68). The Lax-Milgram theorem hence guarantees the existence and uniqueness of a solution. The estimate for the H^1-seminorm of the solution follows directly from (6.60) and (6.65) inserting

$\chi_H = u_H$. The L_2-norm of the solution can be estimated by its H^1-seminorm utilizing the property (6.53) of high-frequency functions. $\square$

A corresponding result holds for the high-order counterpart of the equation (6.68), that formally results from this equation replacing the test function χ_H by test functions $\varepsilon v_H + \mathscr{L} v_H$, with all the care that has to be taken with this type of arguments.

Lemma 6.10. *For frequency bounds Ω as in (6.59), the equation*

$$\tilde{a}(u_H, v_H) + \mu(u_H, \varepsilon v_H + \mathscr{L} v_H) = \tilde{b}(\varphi, v_H), \quad v_H \in X^1_{I,H}, \qquad (6.70)$$

possesses a unique solution $u_H \in X^1_{I,H}$ for all given functions $\varphi \in X^0_I$ and arbitrary nonnegative parameters μ. This solution satisfies the estimate

$$|u_H|_{I,1} \le \Omega \left(\varepsilon \|\varphi\|_0^2 + |\varphi|_{I,0}^2 \right)^{1/2}. \qquad (6.71)$$

Proof. As $\mu \ge 0$ and $(u, \varepsilon u + \mathscr{L} u) \ge 0$ for $u \in X^1_I$, the proposition again follows from the coercivity (6.61) of the bilinear form $\tilde{a}(u_H, v_H)$, from the bound (6.64) for the bilinear form $\tilde{b}(\varphi, v)$ on the right hand side, and the Lax-Milgram theorem. $\square$

We want to show that the solutions of the equations (6.68) and (6.70) coincide for $\varphi \in X^0_I$. For that we need the following, at first sight seemingly obvious lemma:

Lemma 6.11. *The solution $u_H \in X^1_{I,H}$ of the equation (6.70) satisfies the equation (6.68) for all rapidly decreasing functions $\chi_H \in \mathscr{S}_{I,H}$ of the particular form*

$$\chi_H = \varepsilon v_H + \mathscr{L} v_H, \quad v_H \in \mathscr{S}_{I,H}. \qquad (6.72)$$

Proof. It suffices to show that the representation (6.46) holds not only for functions u and v in $\mathscr{D}_I$ but for all functions $u \in X^1_I$ and $v \in \mathscr{S}_I$, and to prove a corresponding relation for the bilinear form (6.62), that, in a strict sense, is defined by (6.62) only for functions φ and v in $\mathscr{D}_I$ and then continuously extended to $X^0_I \times X^1_I$. We begin with the case that $u \in \mathscr{D}_I$ and approximate $v \in \mathscr{S}_I$ by the functions

$$v_R(x) = \phi\left(\frac{x}{R}\right) v(x), \quad R > 0,$$

in $\mathscr{D}_I$, where ϕ is an infinitely differentiable, rotationally symmetric function with values $\phi(x) = 1$ for $|x| \le 1$ and $\phi(x) = 0$ for $|x| \ge 2$. For sufficiently large R, v_R and v coincide on the support of u. As v_R tends to v in the X^1_I-norm, by the definition (6.46) of the bilinear form $\tilde{a}(u, v)$ for functions in $\mathscr{D}_I$

$$\tilde{a}(u, v) = \lim_{R \to \infty} \tilde{a}(u, v_R) = a(u, \varepsilon v + \mathscr{L} v) + \gamma s(u, \varepsilon v + \mathscr{L} v)$$

for all $u \in \mathscr{D}_I$ and $v \in \mathscr{S}_I$. Since the left and the right hand sides of this equation represent bounded linear functionals in $u \in X^1_I$ for $v \in \mathscr{S}_I$ given, and since $\mathscr{D}_I$ is a dense subset of X_I^1, the equation transfers to all $u \in X_I^1$ and $v \in \mathscr{S}_I$. Correspondingly,

$$\widetilde{b}(\varphi,v) = b(\varphi,\varepsilon v + \mathscr{L}v)$$

for all $\varphi \in X_I^0$ and $v \in \mathscr{S}_I$, from which the proposition then follows. □

The argument that closes the gap between the equations (6.68) and (6.70) is the observation that every function in $\chi_H \in \mathscr{S}_{I,H}$ can be represented in the form (6.72). The proof requires that the parameter ε is strictly positive and breaks down for $\varepsilon = 0$.

Lemma 6.12. *For all rapidly decreasing high-frequency functions* $\chi_H \in \mathscr{S}_{I,H}$ *there is a rapidly decreasing high-frequency function* $v_H \in \mathscr{S}_{I,H}$ *that solves the equation*

$$\varepsilon v_H + \mathscr{L}v_H = \chi_H. \tag{6.73}$$

Proof. The antisymmetry of a function with respect to the given permutations transfers to its Fourier transform and vice versa. The function $v_H \in \mathscr{S}_{I,H}$ given by

$$\widehat{v}_H(\omega) = \frac{1}{\varepsilon + \prod_{i \in I}|\omega_i|^2}\,\widehat{\chi}_H(\omega)$$

has by this reason the required symmetry properties and solves the equation. □

The solution of the modified equation (6.70) therefore satisfies the equation (6.68) for all $\chi_H \in \mathscr{S}_{I,H}$ and, as $\mathscr{S}_{I,H}$ is dense in $H_{I,H}^1$, for all $\chi_H \in H_{I,H}^1$. Since the equation (6.68) possesses only one solution, the solutions of both equations coincide for φ in X_I^0 given. Since $\varepsilon > 0$ was arbitrary, this observation and (6.53) prove:

Lemma 6.13. *If the bound* Ω *separating the high from the low frequencies is chosen according to (6.59) and* $\varphi \in X_I^0$, *the solution* $u_H \in H_{I,H}^1$ *of the equation (6.68) is contained in the space* $X_{I,H}^1$ *and satisfies the estimates*

$$|u_H|_{I,0} \leq |\varphi|_{I,0}, \quad |u_H|_{I,1} \leq \Omega|\varphi|_{I,0}. \tag{6.74}$$

Since the low-frequency part u_L of the solution u of the equation (6.11), (6.66) is contained in X_I^0 and even in X_I^1, we can apply the result just proved to the equation (6.67), from which it follows that also the high-frequency part u_H of u and with that u itself are contained in X_I^1. The quantitative version of this result reads:

Theorem 6.7. *The solutions* $u \in H_I^1$ *of the modified eigenvalue problem (6.11) for negative* λ *are contained in* X_I^1. *For frequency bounds (6.59), their seminorms (6.15), (6.19) can be estimated as follows in terms of their low-frequency parts:*

$$|u|_{I,0} \leq \sqrt{2}\,|u_L|_{I,0}, \quad |u|_{I,1} \leq \sqrt{2}\,\Omega\,|u_L|_{I,0}. \tag{6.75}$$

Proof. By Lemma 6.13, the high frequency parts u_H of these u satisfy the estimates

$$|u_H|_{I,0} \leq |u_L|_{I,0}, \quad |u_H|_{I,1} \leq \Omega |u_L|_{I,0}.$$

They can thus be controlled by the corresponding low-frequency parts u_L independent of the given $\lambda < 0$. The proposition follows from the orthogonality of the decomposition into the two parts u_L and u_H and the inverse estimate in (6.54). □

The estimates (6.75) for the mixed derivatives of the solutions have a counterpart for the solutions themselves that follows in the same way directly from Lemma 6.9.

Theorem 6.8. *Under the same assumptions as in Theorem 6.7, the solutions of the modified eigenvalue problem (6.11) satisfy the two estimates*

$$\|u\|_0 \leq \sqrt{2}\|u_L\|_0, \quad |u|_1 \leq \sqrt{2}\,\Omega\,\|u_L\|_0. \tag{6.76}$$

A solution $u \in H_I^1$ of the equation (6.11), (6.66) is trivially contained in $H_{I'}^1$ for all nonempty subsets I' of I. As $s(u,v)$ is obviously invariant under the exchange of all electrons i in the subset I' of I, Theorem 6.2 ensures that u solves the equations

$$a(u,\chi) + \gamma s(u,\chi) = \lambda(u,\chi), \quad \chi \in H_{I'}^1, \tag{6.77}$$

on all of these spaces $H_{I'}^1$ and thus satisfies, by Theorem 6.7, the estimates

$$|u|_{I',0} \leq \sqrt{2}|u_L|_{I',0}, \quad |u|_{I',1} \leq \sqrt{2}\,\Omega\,|u_L|_{I',0} \tag{6.78}$$

for all nonempty subsets I' of the given index set I. Therefore the norms given by

$$\|u\|_{I,1}^2 = \int \left(\sum_{i=1}^N \left|\frac{\omega_i}{\Omega}\right|^2\right) \prod_{i \in I}\left(1 + \left|\frac{\omega_i}{\Omega}\right|^2\right)|\hat{u}(\omega)|^2\,d\omega, \tag{6.79}$$

$$\|u\|_{I,0}^2 = \int \prod_{i \in I}\left(1 + \left|\frac{\omega_i}{\Omega}\right|^2\right)|\hat{u}(\omega)|^2\,d\omega. \tag{6.80}$$

of these functions, that combine the H^1-norm and H^1-norms of the corresponding mixed derivatives, remain finite. The frequency bound Ω fixes a length scale. Such length scales naturally appear in every estimate that relates derivatives of distinct order to each other. They have to be incorporated in the definition of the corresponding norms to compensate the different scaling behavior of the derivatives and to obtain physically meaningful estimates that are independent of the choice of units.

With these notations, we can now formulate and prove our final and conclusive regularity theorem for the solutions of the modified eigenvalue problem (6.11):

Theorem 6.9. *The solutions $u \in H_I^1$ of the modified eigenvalue problem (6.11) for negative values λ satisfy, for frequency bounds (6.59), the estimates*

$$\|u\|_{I,0} \leq \sqrt{2e}\,\|u\|_0, \quad \|u\|_{I,1} \leq \sqrt{2e}\,\|u\|_0. \tag{6.81}$$

Proof. By the estimates (6.76) for the L_2-norm of the solution itself, respectively the estimates (6.78) for the L_2-norms of its corresponding mixed derivatives,

$$\int \prod_{i \in I'} \left| \frac{\omega_i}{\Omega} \right|^2 |\hat{u}(\omega)|^2 \, d\omega \leq 2 \int_{|\omega| \leq \Omega} \prod_{i \in I'} \left| \frac{\omega_i}{\Omega} \right|^2 |\hat{u}(\omega)|^2 \, d\omega \qquad (6.82)$$

for all subsets I' of I, where the empty product is by definition 1. As

$$\sum_{I' \subseteq I} \prod_{i \in I'} \left| \frac{\omega_i}{\Omega} \right|^2 = \prod_{i \in I} \left(1 + \left| \frac{\omega_i}{\Omega} \right|^2 \right), \qquad (6.83)$$

one obtains from (6.82) first the estimate

$$\|u\|_{1,0}^2 \leq 2 \int_{|\omega| \leq \Omega} \prod_{i \in I} \left(1 + \left| \frac{\omega_i}{\Omega} \right|^2 \right) |\hat{u}(\omega)|^2 \, d\omega. \qquad (6.84)$$

The product on the right hand side of (6.83) is, because of

$$\prod_{i \in I} \left(1 + \left| \frac{\omega_i}{\Omega} \right|^2 \right) \leq \exp \left(\sum_{i \in I} \left| \frac{\omega_i}{\Omega} \right|^2 \right), \qquad (6.85)$$

bounded by the constant e for all ω in the ball of radius Ω around the origin. This proves the first of the two estimates. The second is treated in the same way. $\qquad \square$

Theorem 6.9 particularly states that the solutions u of the electronic Schrödinger equation (4.30) itself possess high-order mixed derivatives. Only small portions of the frequency domain substantially contribute to the wave functions. This remark can be quantified with help of the notion of hyperbolic crosses, hyperboloid-like regions in the frequency or momentum-space that consist of those ω for which

$$\prod_{i \in I_-} \left(1 + \left| \frac{\omega_i}{\Omega} \right|^2 \right) + \prod_{i \in I_+} \left(1 + \left| \frac{\omega_i}{\Omega} \right|^2 \right) \leq \frac{1}{\varepsilon^2}, \qquad (6.86)$$

where $\varepsilon > 0$ is a control parameter that determines their size, and I_- and I_+ are again the sets of the indices i of the electrons with spin $\sigma_i = -1/2$ and $\sigma_i = +1/2$ respectively. If u_ε denotes that part of the wave function whose Fourier transform coincides with that of u on this domain and vanishes outside of it, the H^1-error

$$\|u - u_\varepsilon\|_1 = \mathcal{O}(\varepsilon) \qquad (6.87)$$

tends to zero like $\mathcal{O}(\varepsilon)$ with increasing size of the crosses. This observation might serve as a basis for the construction of approximation methods, for example utilizing the fact that functions like the projections u_ε with Fourier transforms vanishing outside such hyperbolic crosses can be sampled on sparse grids [93]. The solutions of the electronic Schrödinger equation in some sense behave like products

$$u(x) = \prod_{i=1}^{N} \phi_i(x_i) \tag{6.88}$$

of orbitals, that is, exponentially decaying functions in H^1, a fact that roughly justifies the picture of atoms and molecules that we have in our minds.

It is remarkable that Theorem 6.9 not only ensures that the given high-order mixed derivatives of the correspondingly exponentially weighted or unweighted eigenfunctions exist and are square integrable, but also gives a rather explicit estimate for their norms in terms of the L_2-norm of the weighted or unweighted eigenfunctions themselves. The estimate (4.11) from Theorem 4.1 implies the lower bound $\lambda \geq -9\theta^2/2$ for the eigenvalues. As $\Sigma(\sigma) \leq 0$, this results in the upper bound

$$\gamma < \sqrt{2(\Sigma(\sigma) - \lambda)} \leq 3\theta(N, Z) \tag{6.89}$$

for the decay rates γ considered in Sect. 6.1. Theorem 6.9 tells us therefore that the estimates (6.81) hold at least for the scaling parameters

$$\Omega \geq (4C + 36)\sqrt{N} \max(N, Z), \tag{6.90}$$

independent of the considered eigenvalue below the ionization threshold, and in particular for the Ω that is equal to the right hand side. There is conversely a minimum

$$\Omega \leq (4C + 36)\sqrt{N} \max(N, Z) \tag{6.91}$$

independent of the choice of the coefficients θ_i in the definition of the exponential weight (provided that the choice of the θ_i maintains the given antisymmetry, of course) such that these estimates hold for all eigenfunctions for these eigenvalues. This minimum Ω can principally be much smaller than the given upper bound and fixes an intrinsic length scale of the considered atomic or molecular system.

6.6 Atoms as Model Systems

The scaling parameter Ω limits the local variation of the wave functions quantitatively. It can be assumed that the right hand side of (6.91) considerably overestimates the optimum Ω for spatially extended molecules that are composed of a big number of light atoms. The question is how sharp this bound is for compact systems with many electrons tightly bound to the nuclei, like heavier atoms. Atoms are, in the given Born-Oppenheimer approximation, described by the Hamilton operator

$$H = \sum_{i=1}^{N} \left\{ -\frac{1}{2}\Delta_i - \frac{Z}{|x_i|} \right\} + \frac{1}{2}\sum_{\substack{i,j=1 \\ i \neq j}}^{N} \frac{1}{|x_i - x_j|}. \tag{6.92}$$

The first term covers the attraction of the electrons by the nucleus and the second their interaction with each other. The crucial property that we utilize here is that the potential in this operator is homogeneous of degree minus one, i.e., that

$$V(\vartheta x) = \vartheta^{-1} V(x) \qquad (6.93)$$

for all $\vartheta > 0$. The H^1-seminorm and the L_2-norm of eigenfunctions of such operators are linked to each other by the famous virial theorem, a proof of which we
include for the sake of completeness. This proof is essentially a reformulation of
that in [86] in terms of weak solutions of the eigenvalue problem.

Theorem 6.10. *The H^1-seminorm and the L_2-norm of an eigenfunction $u \in H^1$ for
the eigenvalue λ of the atomic Hamilton operator (6.92) are linked via the relation*

$$|u|_1^2 = -2\lambda \|u\|_0^2. \qquad (6.94)$$

Proof. Let $u_\vartheta(x) = u(\vartheta x)$ for $\vartheta > 0$. A short calculation only utilizing the fact that
u is an eigenfunction for the eigenvalue λ then shows that

$$\int \nabla u_\vartheta \cdot \nabla v \, dx = 2\vartheta^2 \lambda \int u_\vartheta v \, dx - 2\vartheta^2 \int V(\vartheta x) u_\vartheta v \, dx$$

for arbitrary test functions $v \in H^1$. Because of $V(\vartheta x) = \vartheta^{-1} V(x)$, this reduces to

$$\int \nabla u_\vartheta \cdot \nabla v \, dx = 2\vartheta^2 \lambda \int u_\vartheta v \, dx - 2\vartheta \int V u_\vartheta v \, dx.$$

On the other hand, for all test functions $v \in H^1$,

$$\int \nabla u \cdot \nabla v \, dx = 2\lambda \int uv \, dx - 2 \int V u v \, dx.$$

Setting $v = u$ in the first and $v = u_\vartheta$ in the second case, for $\vartheta \neq 1$ it follows that

$$(\vartheta + 1)\lambda \int u u_\vartheta \, dx = \int V u u_\vartheta \, dx.$$

For all square integrable functions u and v

$$\lim_{\vartheta \to 1} \int v(x) u(\vartheta x) \, dx = \int v(x) u(x) \, dx,$$

as can be shown approximating u by continuous functions with bounded support.
Since for $u \in H^1$ the product Vu is square integrable, too, this yields

$$2\lambda \int u^2 \, dx = \int V u u \, dx.$$

Using once more that u is an eigenfunction, one finally gets the proposition.   □

The virial theorem relates the expectation values of the kinetic energy, the potential
energy, and the total energy to each other, but also determines, through the different

scaling behavior of both sides of the equation, the length scale on which the considered eigenfunction varies. Hence it is no surprise that a lower bound for the optimal scaling parameter Ω can be derived in terms of the eigenvalues.

Theorem 6.11. *If the estimates from Theorem 6.9 hold for the eigenfunction u in $H^1(\sigma)$ for the eigenvalue λ of the atomic Hamilton operator (6.92), necessarily*

$$\Omega \geq \sqrt{\frac{|\lambda|}{e}}. \tag{6.95}$$

Proof. From the virial theorem, from the Fourier representation (6.48) of the H^1-seminorm and of the norm given by (6.79), and from Theorem 6.9 one gets

$$-2\lambda \, \|u\|_0^2 = |u|_1^2 \leq \Omega^2 \, \|u\|_{1,1}^2 \leq 2e\,\Omega^2 \, \|u\|_0^2.$$

Because $u \neq 0$, one can divide by the L_2-norm of u and obtain the proposition. $\square$

Since the ionization threshold $\Sigma(\sigma)$ is less than or equal to zero by Theorem 5.16, the upper estimate resulting from Theorem 6.9 and the lower estimate just derived resulting from the virial theorem lead to the bounds

$$\sqrt{|\Lambda(\sigma)|} \lesssim \Omega \lesssim \sqrt{N}\max(N,Z) + \sqrt{|\Lambda(\sigma)|} \tag{6.96}$$

for the optimum Ω that is independent of the considered eigenvalues $\lambda < \Sigma(\sigma)$. The second term on the right hand side of (6.96) that comes from the additional part (6.9) in the equation (6.8) for the exponentially weighted eigenfunctions will therefore never dominate the asymptotic behavior of the optimum Ω in N and Z.

The problem thus reduces to the question of how well the bound (6.91) reflects the growth of the optimum scaling parameter Ω in N and Z for unweighted eigenfunctions, in which case the second term on the right hand side of (6.91) can be omitted. To answer this question at least partially, we consider the operator

$$H = \sum_{i=1}^{N} \left\{ -\frac{1}{2}\Delta_i - \frac{Z}{|x_i|} \right\} \tag{6.97}$$

in which the electron-electron interaction is completely neglected and to which Theorem 6.11 can be literally transferred. Due to the absence of the electron-electron interaction potential, the estimates (6.81) hold then regardless of any symmetry property. The eigenfunctions of this operator are linear combinations of the products

$$u(x) = \prod_{i=1}^{N} \phi_i(x_i) \tag{6.98}$$

of hydrogen-like wave functions, solutions of the Schrödinger equation

$$-\frac{1}{2}\Delta\phi - \frac{Z}{|x|}\phi = \lambda\phi \tag{6.99}$$

for a single electron in the field of a nucleus of charge Z. The hydrogen-like wave functions are explicitly known and are calculated in almost every textbook on quantum mechanics; see Chap. 9 for details. The corresponding eigenvalues

$$\lambda = -\frac{Z^2}{2n^2}, \quad n = 1, 2, \ldots, \tag{6.100}$$

are highly degenerate. The associated eigenspaces are spanned by the eigenfunctions with the given principal quantum number n, the angular momentum quantum numbers $l = 0, \ldots, n-1$, and the magnetic quantum numbers $m = -l, \ldots, l$ and have dimension n^2. The knowledge about these eigenfunctions forms the basis of our understanding of the periodic table.

If we ignore the Pauli principle, every product (6.98) becomes an admissible eigenfunction. The ground state energy of the corresponding system is then N times the minimum eigenvalue (6.100), i.e., $\lambda = -NZ^2/2$, from which the lower bound

$$\Omega \gtrsim N^{1/2}Z \tag{6.101}$$

follows, which behaves like the upper bound (6.91) in the number N of electrons and the nuclear charge Z for the case of neutral atoms or positively charged ions. Thus neither the upper bound (6.91) nor the lower bound (6.95) can be improved without bringing the Pauli principle or the electron-electron interaction into play.

If the Pauli principle is taken into account, the orbitals ϕ_i in (6.98) have to be partitioned into two groups associated with the electrons with spin up and spin down. The orbitals in each group have to be linearly independent of each other as the product otherwise vanishes under the corresponding antisymmetrization. That increases the ground state energy and correspondingly decreases the lower bound for the scaling parameter. Unlike a real atom, the system attains its minimum energy λ in states in which the numbers of electrons with spin up and spin down differs at most by one, that is, with at most one unpaired electron. Consider, for example, the case that the electrons can be distributed to M doubly occupied shells $n = 1, 2, \ldots, M$ with $2n^2$ electrons in the shell n, n^2 with spin up and n^2 with spin down. Then $\lambda = -MZ^2$. Because $N \sim 2M^3/3$, the minimum eigenvalue hence behaves in the described situation like $\lambda \sim N^{1/3}Z^2$ and the scaling parameter needs therefore to grow at least like

$$\Omega \gtrsim N^{1/6}Z. \tag{6.102}$$

There remains some gap between this lower bound and the upper bound (6.91), but the estimate shows at least that the actual growth of the optimal scaling parameter in N and Z is not substantially overestimated by the right hand side of (6.91) for systems like the ones considered here.

In fact, the observed behavior is not restricted to the model Hamiltonian (6.97). Lieb and Simon [61] proved that the minimum eigenvalue of the full operator (6.92) grows like $\gtrsim Z^{7/3}$ with the nuclear charge Z in the case $Z = N$, i.e., of neutral systems, which confirms the lower estimate (6.102). A more detailed study [94] of the product eigenfunctions (6.98) moreover shows that the optimum Ω behaves in this case indeed like the square root of the ground state energy, which can be explained from the behavior of the orbitals. One may conjecture that this generally holds.

6.7 The Exponential Decay of the Mixed Derivatives

In Sect. 6.5 we have proven that the eigenfunctions themselves as well as the correspondingly exponentially weighted eigenfunctions possess square integrable high-order mixed weak derivatives. In this short concluding section it is shown that the exponentially weighted mixed derivatives of the eigenfunctions are square integrable. This follows essentially from the fact that the corresponding partial derivatives of the exponential weight factors can be estimated by these factors themselves:

Theorem 6.12. *Let $D^\nu u = L_\alpha u$, L_α as in (6.17), be one of the weak partial derivatives of the eigenfunction u whose existence and square integrability follows from the results of Sect. 6.5, and let e^F be one of the associated weight factors for which $D^\nu (e^F u)$ has been shown to be square integrable too. The weighted derivatives*

$$e^F D^\nu u, \quad e^F \frac{\partial}{\partial x_{i,k}} D^\nu u \qquad (6.103)$$

are then square integrable as well.

Proof. The proof is based on the representation

$$D^\nu (e^F u) = \sum_{\mu \le \nu} e^F F_\mu D^{\nu - \mu} u$$

of the corresponding weak derivatives of $e^F u$, that is a generalization of the product rule from Lemma 6.1 and can be derived from it taking into account the special structure of the multi-indices ν considered. The coefficient functions are products

$$F_\mu(x) = \gamma^{|\mu|} \prod_i \theta_i \frac{x_{i,\alpha(i)}}{|x_i|}$$

that run over the components upon which D^μ acts. This representation allows us to express $e^F D^\nu u$ in terms of $D^\nu (e^F u)$ and the weighted lower order derivatives $e^F D^{\nu - \mu} u$ of u. Since the F_μ are uniformly bounded, the square integrability of $e^F D^\nu u$ follows by induction on the order of differentiation. The square integrability of the second function is proven differentiating the representation above. To cover the resulting derivatives of the F_μ one needs again the Hardy inequality.   □

The exponential functions $x \to \exp(F(x))$ dominate every polynomial, regardless the decay rate γ determined by the gap between the considered eigenvalue λ and the ionization threshold. This results in the following corollary of Theorem 6.12:

Theorem 6.13. *Let $D^\nu u = L_\alpha u$, L_α as in (6.17), be one of the weak partial derivatives of the eigenfunction u whose existence and square integrability follows from the results of Sect. 6.5, and let P be an arbitrary polynomial. Then*

$$PD^\nu u, \ P\frac{\partial}{\partial x_{i,k}}D^\nu u \ \in \ L_2. \tag{6.104}$$

This statement can again be reversed. For every multi-index μ the function $D^\nu(x^\mu u)$ and the weighted derivative $\omega^\nu D^\mu \hat{u}$ of its Fourier transform are square integrable. The μ are not subject to restrictions, due to the exponential decay of the wave functions and their mixed derivatives, but the ν are, because of the restricted regularity.

Chapter 7
Eigenfunction Expansions

The aim of this chapter is to derive discrete counterparts of the regularity theorems from Chap. 6 similar to how smoothness can be characterized for periodic functions in terms of the decay rate of their Fourier coefficients. The problem is that the solutions of the electronic Schrödinger equation are defined on the infinitely extended space so that not only their regularity properties but also their decay behavior comes into play and has to be utilized. The foundations for that have been laid in Chap. 6. The idea is to expand the high-dimensional solutions of the Schrödinger equation into series of tensor products of eigenfunctions of three-dimensional operators

$$H = -\Delta + V \tag{7.1}$$

with locally square integrable, nonnegative potentials V for which

$$\lim_{|x| \to \infty} V(x) = +\infty. \tag{7.2}$$

An example of such an operator is the Hamilton operator of the three-dimensional harmonic oscillator studied in detail in Sect. 3.4. As follows from the considerations in Sect. 5.4, the essential spectrum of such operators is empty. They possess an L_2-complete, L_2-orthonormal system of eigenfunctions $\phi_1, \phi_2, \phi_3, \ldots$ for strictly positive eigenvalues $0 < \lambda_1 \le \lambda_2 \le \ldots$ of finite multiplicity. Every square integrable function $u : \mathbb{R}^{3N} \to \mathbb{R}$ can therefore be represented as L_2-convergent series

$$u(x) = \sum_{k \in \mathbb{N}^N} \widehat{u}(k) \prod_{i=1}^{N} \phi_{k_i}(x_i), \quad \widehat{u}(k) = \left(u, \prod_{i=1}^{N} \phi_{k_i} \right), \tag{7.3}$$

where the sum runs over the tensor products of the three-dimensional eigenfunctions, that form an L_2-complete orthonormal system in $L_2(\mathbb{R}^{3N})$. In this chapter we will examine the convergence properties of such series in H^1 for the eigenfunctions of the electronic Schrödinger operator that obey the Pauli principle. It will turn out that only very few of the products substantially contribute to these eigenfunctions.

H. Yserentant, *Regularity and Approximability of Electronic Wave Functions*,
Lecture Notes in Mathematics 2000, DOI 10.1007/978-3-642-12248-4_7,
© Springer-Verlag Berlin Heidelberg 2010

7.1 Discrete Regularity

As in the previous chapters we fix the spin distribution of the electrons and denote by I_- and I_+ the sets of the indices $1, \ldots, N$ of the electrons with spin $\sigma_i = -1/2$ and $\sigma_i = +1/2$ respectively. The aim is to show that the norm given by the expression

$$\|u\|^2 = \sum_k \left(\sum_{i=1}^N \frac{\lambda_{k_i}}{\Omega^2} \right) \left(\prod_{i \in I_-} \frac{\lambda_{k_i}}{\Omega^2} + \prod_{i \in I_+} \frac{\lambda_{k_i}}{\Omega^2} \right) |\widehat{u}(k)|^2 \qquad (7.4)$$

of the solutions of the electronic Schrödinger equation for eigenvalues below the ionization threshold remains bounded by a weighted L_2-norm; the constant Ω is here the same as in Chap. 6 and fixes the length scale on which the solutions vary.

We consider again the two parts of the norm separately and select one of the index sets I_- and I_+ that we denote by I. The first step is to rewrite the given part of the norm in terms of the differential operators

$$H_i = -\Delta_i + V_i, \quad V_i(x) = V(x_i), \qquad (7.5)$$

that act upon the coordinates of the electron i. We first restrict ourselves hereby to the functions in $\mathscr{D}$, the infinitely differentiable functions with bounded support.

Lemma 7.1. *For all infinitely differentiable functions u and v with bounded support,*

$$\sum_k \left(\sum_{j=1}^N \lambda_{k_j} \right) \left(\prod_{i \in I} \lambda_{k_i} \right) \widehat{u}(k) \widehat{v}(k) = \sum_{j=1}^N B(u, H_j v), \qquad (7.6)$$

where the bilinear form on the right hand side is given by

$$B(u, v) = \left(\left(\prod_{i \in I} H_i \right) u, v \right). \qquad (7.7)$$

Proof. Expanding the functions $H_j v \in L_2$, we first obtain the representation

$$\sum_{j=1}^N B(u, H_j v) = \sum_k \left(\left(\prod_{i \in I} H_i \right) u, \prod_{i=1}^N \phi_{k_i} \right) \left(\left(\sum_{j=1}^N H_j \right) v, \prod_{i=1}^N \phi_{k_i} \right).$$

From the weak form of the three-dimensional eigenvalue problem one obtains

$$\left(H_j w, \prod_{i=1}^N \phi_{k_i} \right) = \lambda_{k_j} \left(w, \prod_{i=1}^N \phi_{k_i} \right),$$

for square integrable functions w with compact support that are infinitely differentiable with respect to the given x_j, and from that, step by step, the proposition. $\square$

The next step is to split the terms on the right hand side of (7.6) into terms that can be estimated separately with help of the results and estimates from Chap. 6.

Lemma 7.2. *For all infinitely differentiable functions u with bounded support,*

$$B(u, H_j u) \leq 2 \sum_{I_1, I_2} (-1)^{|I_2|} \left(\left(\prod_{i \in I_1} V_i \right) \left(-\Delta_j \prod_{i \in I_2} \Delta_i \right) u, u \right) \tag{7.8}$$

$$+ 2 \sum_{I_1, I_2} (-1)^{|I_2|} \left(\left(V_j \prod_{i \in I_1} V_i \right) \left(\prod_{i \in I_2} \Delta_i \right) u, u \right),$$

where the first sum on the right hand side of this estimate extends over all partitions $I = I_1 \cup I_2$ of the index set I into disjoint subsets I_1 and I_2 for which $j \notin I_1$, and the second sum over all partitions $I = I_1 \cup I_2$ of I for which $j \notin I_2$.

Proof. The idea is to split the product of the H_i into a sum of products of the Δ_i and and of the V_i and to utilize that Δ_i and V_j commute for $i \neq j$. For indices $j \in I$ first

$$B(u, H_j u) = \sum_{I'_1, I'_2} (-1)^{|I'_2|} \left(\left(\prod_{i \in I'_1} V_i \right) \left(\prod_{i \in I'_2} \Delta_i \right) (-\Delta_j + V_j) u, (-\Delta_j + V_j) u \right),$$

where the sum runs over the disjoint partitions $I' = I'_1 \cup I'_2$ of $I' = I \setminus \{j\}$. Since

$$\langle v, w \rangle := (-1)^{|I'_2|} \left(\left(\prod_{i \in I'_1} V_i \right) \left(\prod_{i \in I'_2} \Delta_i \right) v, w \right),$$

is a symmetric, positive semidefinite bilinear form for I'_1 and I'_2 given and thus

$$\langle v+w, v+w \rangle \leq 2 \langle v, v \rangle + 2 \langle w, w \rangle,$$

one obtains from this relation, setting $v = -\Delta_j u$ and $w = V_j u$, by integration by parts

$$B(u, H_j u) \leq 2 \sum_{I'_1, I'_2} (-1)^{|I'_2|+1} \left(\left(\prod_{i \in I'_1} V_i \right) \left(-\Delta_j \prod_{i \in I'_2 \cup \{j\}} \Delta_i \right) u, u \right)$$

$$+ 2 \sum_{I'_1, I'_2} (-1)^{|I'_2|} \left(\left(V_j \prod_{i \in I'_1 \cup \{j\}} V_i \right) \left(\prod_{i \in I'_2} \Delta_i \right) u, u \right).$$

This is obviously an estimate of the form (7.8). For the indices $j \notin I$ simply

$$B(u, H_j u) = \sum_{I_1, I_2} (-1)^{|I_2|} \left(\left(\prod_{i \in I_1} V_i \right) \left(-\Delta_j \prod_{i \in I_2} \Delta_i \right) u, u \right)$$

$$+ \sum_{I_1, I_2} (-1)^{|I_2|} \left(\left(V_j \prod_{i \in I_1} V_i \right) \left(\prod_{i \in I_2} \Delta_i \right) u, u \right),$$

where the sums run over the partitions of I into disjoint subsets I_1 and I_2. As the terms of which the sums are composed are all nonnegative, (7.8) follows.   □

To keep control of the right hand side of (7.8), one needs to bound the growth of the potentials V_i in (7.5) respectively that of the original potential V. Let

$$V_i^*(x) = \frac{\Lambda_0}{R} \exp\left(\left|\frac{x_i}{R}\right|\right), \tag{7.9}$$

with Λ_0 a given constant and R a scaling parameter to be discussed later. We assume

$$V_i(x) \le V_i^*(x)^2. \tag{7.10}$$

This condition holds automatically for polynomially growing potentials V, as for the case of the harmonic oscillator, independent of the choice of the scaling parameter R and be it at the price of a large constant Λ_0. The assumption (7.10) allows us to estimate the right hand sides in (7.8) and with that the left hand side of (7.6) in terms of the norms (6.15) of the correspondingly exponentially weighted functions.

Lemma 7.3. *For all infinitely differentiable functions u with bounded support*

$$\sum_k \left(\sum_{j=1}^N \frac{\lambda_{k_j}}{\Omega^2}\right)\left(\prod_{i\in I} \frac{\lambda_{k_i}}{\Omega^2}\right)|\hat{u}(k)|^2 \le 2 \sum_{I_1,I_2}\left(\frac{1}{\Omega^2}\right)^{|I_2|+1}\left|\left(\prod_{i\in I_1}\frac{V_i^*}{\Omega}\right)u\right|_{I_2,1}^2 \tag{7.11}$$

$$+ 2 \sum_{j=1}^N \sum_{I_1,I_2}\left(\frac{1}{\Omega^2}\right)^{|I_2|}\left|\left(\frac{V_j^*}{\Omega}\prod_{i\in I_1}\frac{V_i^*}{\Omega}\right)u\right|_{I_2,0}^2,$$

where the sums run over the partitions $I_1 \cup I_2$ of I and $j \notin I_2$ in the second case.

Proof. The products of the Laplacians in (7.8) can, as in (6.18), be written as sums of squares of operators. These operators commute with the corresponding multiplication operators and can be distributed in equal parts to both sides of the inner products. In the notion introduced in Sect. 6.2, integration by parts leads to

$$(-1)^{|I_2|}\left(\left(\prod_{i\in I_1} V_i\right)\left(-\Delta_j \prod_{i\in I_2} \Delta_i\right)u, u\right) = \sum_{\alpha\in I_2^*}\left(\left(\prod_{i\in I_1} V_i\right)\nabla_j L_\alpha u, \nabla_j L_\alpha u\right)$$

as I_1 and I_2 are disjoint and $j \notin I_1$. The other terms on the right hand side of (7.8) are treated correspondingly utilizing $j \notin I_2$ and

$$(-1)^{|I_2|}\left(\left(V_j \prod_{i\in I_1} V_i\right)\left(\prod_{i\in I_2} \Delta_i\right)u, u\right) = \sum_{\alpha\in I_2^*}\left(\left(V_j \prod_{i\in I_1} V_i\right)L_\alpha u, L_\alpha u\right).$$

The proposition thus follows from Lemmas 7.1 and 7.2, the non-negativity of the potentials V_i, and the assumption (7.10) on their growth.   □

As we know from Chap. 6, the seminorms on the right hand side of (7.11) of an eigenfunction $u \in H^1(\sigma)$ for an eigenvalue λ below the ionization threshold $\Sigma(\sigma)$ remain finite as long as the constant R is chosen sufficiently large in dependence on the gap between the eigenvalue and the ionization threshold. Theorems 6.7 and 6.8 and the inverse estimate (6.52) for the low-frequency parts of functions yield then

$$\left(\frac{1}{\Omega^2}\right)^{|I_2|} \left\|\left(\frac{V_j^*}{\Omega} \prod_{i \in I_1} \frac{V_i^*}{\Omega}\right) u \right\|_{I_2,0}^2 \leq 2 \left\|\left(\frac{V_j^*}{\Omega} \prod_{i \in I_1} \frac{V_i^*}{\Omega}\right) u \right\|_0^2, \tag{7.12}$$

$$\left(\frac{1}{\Omega^2}\right)^{|I_2|+1} \left\|\left(\prod_{i \in I_1} \frac{V_i^*}{\Omega}\right) u \right\|_{I_2,1}^2 \leq 2 \left\|\left(\prod_{i \in I_1} \frac{V_i^*}{\Omega}\right) u \right\|_0^2. \tag{7.13}$$

The final step is therefore essentially to transfer the estimate (7.11) to the classes of functions to which the solutions of the electronic Schrödinger equation belong.

Lemma 7.4. *A square integrable function $u : \mathbb{R}^{3N} \to \mathbb{R}$ that possesses square integrable weak derivatives of corresponding orders and for which the expressions*

$$\left\|\left(\prod_{i \in I_1} \frac{V_i^*}{\Omega}\right) u \right\|_{J,1}^2, \quad \left\|\left(\frac{V_j^*}{\Omega} \prod_{i \in I_1} \frac{V_i^*}{\Omega}\right) u \right\|_{J,0}^2, \tag{7.14}$$

remain finite for all disjoint index sets $I_1, J \subseteq I$ and all indices $j \notin J$ can, in the sense of the norm induced by the right hand side of (7.11), be approximated arbitrarily well by functions in $\mathcal{D}$ and is thus contained in the completion of $\mathcal{D}$ under this norm.

Proof. Let χ be an infinitely differentiable cut-off function with values $\chi(x) = 1$ for $|x| \leq 1$ and $\chi(x) = 0$ for $|x| \geq 2$ and set $\chi_\vartheta(x) = \chi(x/\vartheta)$. The functions $u_\vartheta = \chi_\vartheta u$ then possess weak derivatives of all considered orders. Moreover,

$$\lim_{\vartheta \to \infty} \|u - u_\vartheta\| = 0$$

in the mentioned norm. This follows from the dominated convergence theorem, because the functions χ_ϑ are uniformly bounded and tend pointwise to one and because their derivatives tend uniformly to zero for ϑ tending to infinity. Thus it suffices to approximate the functions u_ϑ. But this is possible without difficulties since the V_i^* and their involved derivatives remain bounded on bounded sets.   □

The finite parts of the left hand side of (7.11), and with that also the complete left hand side, can therefore be estimated by the right hand side of this equation for the eigenfunctions u of the electronic Schrödinger operator. This yields:

Theorem 7.1. *Provided the potentials V_i satisfy the estimate (7.10) with R chosen sufficiently large in dependence on the gap between the considered eigenvalue and the ionization threshold, the given eigenfunctions u satisfy the estimate*

$$\sum_k \left(\sum_{i=1}^N \frac{\lambda_{k_i}}{\Omega^2}\right)\left(\prod_{i \in I_-} \frac{\lambda_{k_i}}{\Omega^2} + \prod_{i \in I_+} \frac{\lambda_{k_i}}{\Omega^2}\right) |\hat{u}(k)|^2 \leq 4(u, Wu), \tag{7.15}$$

where the weight function $W = W_- + W_+$ is composed of the two parts

$$W_\pm = \left(1 + \sum_{i=1}^{N} \left|\frac{V_i^*}{\Omega}\right|^2\right) \prod_{i \in I_\pm} \left(1 + \left|\frac{V_i^*}{\Omega}\right|^2\right). \tag{7.16}$$

Proof. It remains only to estimate the terms on the right hand side of (7.11) as in (7.12) and (7.13) and to note that the single terms can, because of the identity

$$\sum_{I_1 \subseteq I_\pm} \left(\prod_{i \in I_1} \frac{V_i^*}{\Omega}\right)^2 = \prod_{i \in I_\pm} \left(1 + \left|\frac{V_i^*}{\Omega}\right|^2\right),$$

be combined into the right hand side of (7.15) to finish the proof of the theorem.   □

Theorem 7.1 is the central result of this chapter and measures the regularity of the considered solutions of the electronic Schrödinger equation in terms of the decay rate of their expansion coefficients. It is interesting to note that the right hand side of (7.15) no longer contains derivatives. The weights (7.16) are of the same structure as the weights in the norms (6.79) and (6.80) that have been considered in Chap. 6, but here in the position space and not in the Fourier space. The theorem shows that the question whether the expressions (7.4) are bounded or not solely depends on the decay behavior of the solutions. The size of their derivatives enters only indirectly via the constant Ω that measures their variation. The decay rate again depends on the gap between the considered eigenvalues of the Schrödinger operator and the ionization threshold. This gap determines, via the relation (7.10), the admissible operators (7.1) and with that their eigenfunctions ϕ_k and their eigenvalues λ_k, whose growth finally determines the speed of convergence of the expansion.

Theorem 7.1 offers a lot of freedom in the choice of the potentials V in the three-dimensional operator (7.1) on which the whole construction is based. The most obvious possibility is to start from a three-dimensional reference potential

$$V_0(x) \le \{\Lambda_0 \exp(|x|)\}^2 \tag{7.17}$$

that is independent of the considered solutions of the equation, and to set

$$V(x) = \frac{1}{R^2} V_0\left(\frac{x}{R}\right). \tag{7.18}$$

The eigenfunctions and eigenvalues of the operator (7.1) are linked to the eigenfunctions $\phi_k^{(0)}$ and eigenvalues $\lambda_k^{(0)}$ of the reference operator $-\Delta + V_0$ by the relation

$$\phi_k(x) = \frac{1}{R^{3/2}} \phi_k^{(0)}\left(\frac{x}{R}\right), \qquad \lambda_k = \frac{\lambda_k^{(0)}}{R^2}, \tag{7.19}$$

that is, by a rescaling. The product ΩR that then appears on both sides of the estimate (7.15) relates the length scale R, that measures the extension of the system, to the length scale $1/\Omega$, on which the considered solutions vary. After renormalization of u, both the norm (7.4) and (u, Wu) are then invariant under a change of units,

$$R \to \vartheta R, \quad \Omega \to \vartheta^{-1}\Omega, \tag{7.20}$$

or correspondingly $x \to \vartheta^{-1}x$ and $\omega \to \vartheta\,\omega$, in the position, respectively in the momentum or Fourier space. The estimate (7.15) becomes invariant under such a change of units and depends only on the dimensionless ratio ΩR of the length scales R and $1/\Omega$ measuring the oscillatory behavior of the considered solutions u of the Schrödinger equation, but not on these quantities themselves.

7.2 Antisymmetry

The physically admissible solutions of the electronic Schrödinger equation are anti-symmetric under the permutation of the electrons with the same spin. This property is reflected in their expansion (7.3) into the product of the three-dimensional eigen-functions. Let G denote the group of these permutations and let

$$(\mathscr{A}v)(x) = \frac{1}{|G|}\sum_{P\in G}\text{sign}(P)v(Px) \tag{7.21}$$

be the corresponding antisymmetrization operator. It reproduces functions in the given solution space $H^1(\sigma)$ and in the associated space $L_2(\sigma)$ of square integrable functions. The operator $\mathscr{A}$ is symmetric with respect to the L_2-inner product and bounded as linear operator from L_2 to L_2. Introducing the notation

$$\psi(k,x) = \prod_{i=1}^{N}\phi_{k_i}(x_i). \tag{7.22}$$

for the tensor products of the three-dimensional eigenfunctions, for $u \in L_2(\sigma)$ thus

$$u(x) = \sum_{k\in\mathbb{N}^N}\widehat{u}(k)\,\mathscr{A}\,\psi(k,x). \tag{7.23}$$

Since $\psi(Qk,x) = \psi(k,Q^{-1}x)$ and because of the group properties of the considered set of permutations, the antisymmetrized basis functions transform like

$$\mathscr{A}\,\psi(Qk,x) = \text{sign}(Q)\,\mathscr{A}\,\psi(k,x) \tag{7.24}$$

under the given permutations of the multi-indices k. They vanish when two entries of k associated with electrons of the same spin coincide. Since for $u \in L_2(\sigma)$,

$$\widehat{u}(k) = (u, \psi(k,\cdot)) = (\mathscr{A}u, \psi(k,\cdot)) = (u, \mathscr{A}\psi(k,\cdot)), \tag{7.25}$$

one can combine the remaining terms for which the multi-indices coincide up to one of the given permutations of the indices. Introducing the antisymmetrized and renormalized, pairwise orthogonal tensor product basis functions

$$\widetilde{\psi}(k,x) = \sqrt{|G|}\,\mathscr{A}\,\psi(k,x) \tag{7.26}$$

that can be written as product of two determinants, or as single determinant when all electrons have the same spin, the expansion of functions $u \in L_2(\sigma)$ into the tensor products of the given eigenfunctions reduces to the orthogonal decomposition

$$u(x) = \sum_k \widetilde{u}(k)\widetilde{\psi}(k,x), \quad \widetilde{u}(k) = (u, \widetilde{\psi}(k, \cdot)), \tag{7.27}$$

where k runs over a set of representatives and those k can be excluded for which two entries k_i associated with electrons of the same spin coincide. The estimate (7.15) from Theorem 7.1 for the corresponding eigenfunctions u of the electronic Schrödinger operator transfers in the given circumstances to the estimate

$$\sum_k \left(\sum_{i=1}^{N} \frac{\lambda_{k_i}}{\Omega^2} \right) \left(\prod_{i\in I_-} \frac{\lambda_{k_i}}{\Omega^2} + \prod_{i\in I_+} \frac{\lambda_{k_i}}{\Omega^2} \right) |\widetilde{u}(k)|^2 \leq 4\,(u, Wu), \tag{7.28}$$

where the sum extends over the same small subset of the multi-indices k as in (7.27) and the size of the single expansion coefficients increases correspondingly. The standard situation is that the indices $i = 1, \ldots, N_-$ label the N_- electrons with spin $-1/2$ and the indices $i = N_- + 1, \ldots, N$ the $N_+ = N - N_-$ electrons with spin $+1/2$. A possible set of representatives consists then of the multi-indices k with components

$$k_1 > \ldots > k_{N_-}, \quad k_{N_-+1} > \ldots > k_N. \tag{7.29}$$

The symmetry group G consists then of $|G| = N_-!\,N_+!$ elements, the factor by which the number of the basis functions diminishes through antisymmetrization.

7.3 Hyperbolic Cross Spaces

Theorem 7.1 states that only a very small part of the terms in the expansion (7.3) of a solution of the electronic Schrödinger equation makes a substantial contribution. Consider the finite dimensional space that is spanned by the tensor products (7.22), respectively by their antisymmetrized counterparts considered in the previous section, for which the associated eigenvalues λ_{k_i} satisfy an estimate of the form

$$\prod_{i\in I_-} \frac{\lambda_{k_i}}{\Omega^2} + \prod_{i\in I_+} \frac{\lambda_{k_i}}{\Omega^2} < \frac{1}{\varepsilon^2}, \tag{7.30}$$

or, which is because $2ab \le a^2 + b^2$ slightly less restrictive, an estimate of the form

$$\left\{ \prod_{i=1}^{N} \frac{\lambda_{k_i}}{\Omega^2} \right\}^{1/2} < \frac{1}{2\varepsilon^2}. \tag{7.31}$$

Due to the obvious geometrical meaning of the products such spaces of ansatz functions are denoted as hyperbolic cross spaces. Their dimensions are much smaller than those of the common spaces that can be associated with balls of the form

$$\sum_{i=1}^{N} \frac{\lambda_{k_i}}{\Omega^2} < \frac{1}{\varepsilon^2} \tag{7.32}$$

and have comparable approximation properties. Their use goes back to the Russian school of numerical analysis [7, 53, 54, 76]. The sparse grid spaces [15] that originated from the work of Zenger [97] are based on the same kind of ideas. They are meanwhile very popular in the treatment of higher-dimensional problems.

Let u_ε be the L_2-orthogonal projection of one of the solutions u of the electronic Schrödinger equation to which Theorem 7.1 applies onto such a hyperbolic cross space that is determined by the conditions (7.30) or (7.31). Moreover, let

$$\|u\|^2 = \sum_{k} \left(\sum_{i=1}^{N} \frac{\lambda_{k_i}}{\Omega^2} \right) |\hat{u}(k)|^2. \tag{7.33}$$

Since u_ε is the part of the expansion (7.3) of u associated with the selected product functions (7.22), respectively the eigenvalues λ_{k_i} for which (7.30) or (7.31) hold,

$$\|u - u_\varepsilon\| \le \varepsilon \||u - u_\varepsilon|\| \le \varepsilon \||u|\|. \tag{7.34}$$

As the norm given by (7.4) dominates the H^1-norm up to a rather harmless constant, u_ε approximates the solution with an H^1-error of order ε if one lets the parameter ε determining the size of the hyperbolic crosses tend to zero. The speed of convergence is determined by the speed with which the eigenvalues λ_k of the underlying three-dimensional operator tend to infinity. For sufficiently fast increasing potentials they grow rapidly as will be shown in the next chapter. Hence a surprisingly high convergence rate, related to the space dimension $3N$, can be achieved, at least if one takes the antisymmetry of the wave functions into account as described in Sect. 7.2.

Chapter 8
Convergence Rates and Complexity Bounds

We have seen in the previous chapter that the expansion of a solution of the $3N$-dimensional electronic Schrödinger equation for eigenvalues below the ionization threshold into correspondingly antisymmetrized products of eigenfunctions of three-dimensional Schrödinger-like operators (7.1) with sufficiently fast increasing potentials converges very rapidly, provided that the three-dimensional eigenvalues tend sufficiently fast to infinity. This chapter is devoted to the quantitative study of this convergence behavior. We begin in Sect. 8.1 with the examination of the growth of the three-dimensional eigenvalues λ_k and show that they increase like

$$\lambda_k \gtrsim k^{\alpha/3}, \tag{8.1}$$

under conditions that are easy to fulfill. The three comes from the fact that we start from an expansion into products of three-dimensional eigenfunctions. The constant α is related to the growth of the underlying potential. For the Hamiltonian of the three-dimensional harmonic oscillator, that falls into the considered class, this constant takes the value $\alpha = 1$. Every value $\alpha < 2$ can be reached with sufficiently rapidly increasing potentials, but not the value $\alpha = 2$ itself. That is, the products in the estimates (7.15) respectively (7.28) and the definition of the hyperbolic cross spaces from Sect. 7.3 increase like powers of factorials. Hence it remains to estimate the number of sequences of positive integers $k_1 > \ldots > k_N$ satisfying an estimate

$$\prod_{i=1}^{N} k_i \leq 2^L, \tag{8.2}$$

where N is here the number of electrons with spin $-1/2$ and $+1/2$ respectively and 2^L is a bound determining the accuracy. This is a number theoretic problem. We will give estimates for the number of these sequences. In particular we will show that their number remains bounded independent of the number of the electrons and essentially grows like $\sim 2^L$ as L tends to infinity. This means that the rate of convergence, measured in the number of the involved antisymmetrized products or determinants, does not deteriorate with the number of electrons.

H. Yserentant, *Regularity and Approximability of Electronic Wave Functions*,
Lecture Notes in Mathematics 2000, DOI 10.1007/978-3-642-12248-4_8,
© Springer-Verlag Berlin Heidelberg 2010

8.1   The Growth of the Eigenvalues in the 3d-Case

The study of the growth of the eigenvalues of second-order elliptic differential operators is a classical topic. It has a long history that began with the work of Weyl [89] and Courant and Hilbert [21] in the first third of the last century. The growth of the eigenvalues of three-dimensional Schrödinger operators (7.1) is examined in [81] and, for the case of rotationally symmetric potentials, in great detail in [82,83], and [84]. In this section, we derive some simple, but for our purposes sufficient estimates for the growth of the eigenvalues of operators of the form (7.1), (7.2). We begin with the example of the three-dimensional harmonic oscillator

$$H\phi = -\Delta\phi + \omega^2 |x|^2 \phi. \tag{8.3}$$

The eigenfunctions and eigenvalues of this operator have been calculated in Sect. 3.4. The eigenfunctions are products of rescaled Hermite polynomials (or linear combinations of such products) with a fixed Gaussian and the eigenvalues read

$$\lambda_k = (2n+3)\omega, \quad n = 0,1,2,\ldots. \tag{8.4}$$

The eigenspaces for these eigenvalues are highly degenerate and have the dimension

$$\frac{(n+1)(n+2)}{2}, \tag{8.5}$$

which is the number of the possible representations of n as a sum $n = n_1 + n_2 + n_3$ of three nonnegative integers n_1, n_2, and n_3 in given order. From this we obtain:

Lemma 8.1. *The ascendingly ordered eigenvalues (8.4) of the three-dimensional harmonic oscillator, counted with multiplicities, satisfy the lower estimate*

$$\lambda_k \geq 3\omega k^{1/3} \tag{8.6}$$

and behave asymptotically like $\lambda_k \sim (48k)^{1/3}\omega$ *for k tending to infinity.*

Proof. By (8.4) and (8.5), $\lambda_k = (2n+3)\omega$ for $n \geq 1$ if and only if

$$\sum_{\ell=0}^{n-1} \frac{(\ell+1)(\ell+2)}{2} < k \leq \sum_{\ell=0}^{n} \frac{(\ell+1)(\ell+2)}{2}.$$

The estimate (8.6) therefore holds for all $k \geq 2$ because, for all $n \geq 1$,

$$\sum_{\ell=0}^{n} \frac{(\ell+1)(\ell+2)}{2} \leq \left(\frac{2n+3}{3}\right)^3,$$

and remains true for $k = 1$. The asymptotic representation of the eigenvalues follows from the fact that both sums behave like $\sim n^3/6$ for n tending to infinity.   □

That this growth is not the best possible is suggested by the example of the eigenvalues of the Laplace operator on a cube, that is, by the eigenvalue problem

$$-\Delta\phi = \lambda\phi, \quad \phi|\partial Q = 0, \tag{8.7}$$

on the region $Q = (0, \pi R)^3$. Solutions of this eigenvalue problem are

$$\phi_k(x) = \left(\frac{2}{\pi R}\right)^{3/2} \prod_{i=1}^{3} \sin\left(n_i \frac{x_i}{R}\right), \quad \lambda_k = \frac{n_1^2 + n_2^2 + n_3^2}{R^2}, \tag{8.8}$$

where n_1, n_2, and n_3 are now natural numbers. As one can again associate exactly one eigenfunction to every such ordered triple of natural numbers, one gets:

Lemma 8.2. *The ascendingly ordered eigenvalues of the Laplace operator from (8.8), again counted with multiplicities, can be estimated from above as*

$$\lambda_k \leq 12R^{-2}k^{2/3}. \tag{8.9}$$

Proof. We assign to the triples (n_1, n_2, n_3) the axiparallel cubes of side length 1 with these triples as upper right corners. The number of the triples for which

$$n_1^2 + n_2^2 + n_3^2 \leq 3L^2$$

is then equal to the total volume of the assigned cubes. Since these cubes cover a cube of side length $\lfloor L \rfloor$, their total volume and with that the number of these triples is at least $(L-1)^3$. The proposition follows choosing $(L-1)^3 = k$. □

A more detailed analysis shows that the above system of eigenfunctions is complete and that the eigenvalues of the negative Laplace operator indeed grow like $\sim k^{2/3}$. Not much surprisingly, the eigenvalues of operators of the form (7.1), (7.2) cannot grow faster than those. Even worse:

Theorem 8.1. *The eigenvalues $\lambda_1 \leq \lambda_2 \leq \ldots$ of an operator of the given form with a continuous potential V tend toward infinity slower than $\sim k^{2/3}$ in the sense that*

$$\lim_{k\to\infty} \frac{\lambda_k}{k^{2/3}} = 0. \tag{8.10}$$

Proof. We begin with the observation that the eigenfunctions ϕ_k of the negative Laplace operator from (8.8) can be extended by the value zero to functions in $H^1(\mathbb{R}^3)$, a fact that can be easily checked by direct calculation going back to the definition of weak derivatives. Let $\mathcal{V}_k$ be the k-dimensional subspace of $H^1(\mathbb{R}^3)$ that is spanned by the eigenfunctions $\phi_1, \ldots, \phi_k$ and assume that the assigned eigenvalues $\lambda_1', \ldots, \lambda_k'$ are ascendingly ordered. As $(\phi_i, \phi_j) = \delta_{ij}$ and $(\nabla\phi_i, \nabla\phi_j) = \lambda_i'\delta_{ij}$ then

$$(\nabla\phi, \nabla\phi) + (V\phi, \phi) \leq (\nabla\phi, \nabla\phi) + M(R)(\phi, \phi) \leq \lambda_k' + M(R)$$

for all functions $\phi \in \mathcal{V}_k$ with L_2-norm one, where $M(R)$ denotes the maximum of the function V on the given cube Q of side length πR. By the min-max characterization of the eigenvalues of H from Theorem 5.9 therefore

$$\lambda_k \leq \lambda'_k + M(R),$$

or, if we insert the upper estimate from Lemma 8.2,

$$\lambda_k \leq 12 R^{-2} k^{2/3} + M(R)$$

for $R > 0$ arbitrary. Since $R \to R^2 M(R)$ is a continuous function that increases monotonely from zero to infinity, there is a minimum $R = R(k)$ for which it attains the value $R^2 M(R) = k^{2/3}$. If we insert this particular R into our estimate, we obtain

$$\lambda_k \leq 13 R(k)^{-2} k^{2/3}.$$

Every computable lower bound for the quantities $R(k)$, and particularly every strictly monotone increasing function $M^* \geq M$ for which the solution of the equation $R^2 M^*(R) = k^{2/3}$ can be explicitly given, thus leads to an upper bound for the eigenvalues. Since $R(k)$ tends in any case to infinity for k tending to infinity, the estimate proves the proposition. $\qquad\qquad\qquad\qquad\qquad\qquad\qquad\qquad\qquad\qquad\qquad\quad\Box$

The result transfers to all potentials that are bounded from above by a continuous potential tending to infinity. This fact limits the order of convergence that one can reach with such expansions into tensor products of three-dimensional eigenfunctions. It is, however, possible to approach the growth $\sim k^{2/3}$ arbitrarily.

Theorem 8.2. *If the potential V can be estimated from below as*

$$V(x) \geq \kappa |x|^\beta \tag{8.11}$$

with $\kappa > 0$ and $\beta \geq 2$, the eigenvalues grow at least like

$$\lambda_k \geq c k^{\alpha/3}, \quad \alpha = \frac{2\beta}{\beta + 2}, \tag{8.12}$$

where c is a positive constant that depends only on κ and β.

Proof. We first assume $\beta > 2$. An simple calculation shows then that there is a constant $a > 0$ that depends on κ and β, but is independent of ω, such that

$$\kappa r^\beta \geq \omega^2 r^2 - a \omega^p, \quad p = \frac{2\beta}{\beta - 2},$$

holds for all $r \geq 0$. If we denote by $\lambda'_1 \leq \lambda'_2 \leq \ldots$ the eigenvalues of the harmonic oscillator (8.3), the min-max characterization of the eigenvalues yields

$$\lambda_k \geq \lambda_k' - a\omega^p$$

or, with the lower estimate from Lemma 8.1 for these eigenvalues,

$$\lambda_k \geq 3\omega k^{1/3} - a\omega^p.$$

If one maximizes the right hand side with respect to ω, one obtains (8.12). The case $\beta = 2$ can, with help of the min-max characterization of the eigenvalues and Lemma 8.1, be directly reduced to the case of the harmonic oscillator.   □

We remark that one can, with the technique from the proof of Theorem 8.1, easily show that from the reverse estimate

$$V(x) \leq \kappa'|x|^\beta \tag{8.13}$$

for the potential in (7.1) conversely a lower bound

$$\lambda_k \leq c'k^{\alpha/3} \tag{8.14}$$

with the same α as in (8.12) follows. Theorem 8.2 therefore yields the correct exponent for the case that the potential can be enclosed between two such bounds.

One can even go further and consider exponentially growing potentials. Such potentials fully exhaust the possible growth, as follows directly from Theorem 8.2:

Theorem 8.3. *If the potential V grows faster than any polynomial in the sense that there exists, for every $\beta \geq 2$, a constant $\kappa = \kappa(\beta) > 0$ such that (8.11) holds,*

$$\lim_{k\to\infty} \frac{k^{\alpha/3}}{\lambda_k} = 0 \tag{8.15}$$

for all exponents α in the interval $0 < \alpha < 2$.

8.2 A Dimension Estimate for Hyperbolic Cross Spaces

We have shown in the previous section that, for a proper choice of the underlying three-dimensional operator, its eigenvalues λ_k increase like $\gtrsim k^{\alpha/3}$, where α can approach the value 2 arbitrarily but cannot reach it. Hence it remains to estimate the number of sequences $k_1 > \ldots > k_N$ of natural numbers, where N here denotes the number of electrons with spin $-1/2$ and spin $+1/2$ respectively, for which

$$\prod_{i=1}^N k_i \leq 2^L. \tag{8.16}$$

The minimum value that this product can attain is $N!$, so that its size at least partly counterbalances the size of the other quantities for bigger N. The problem to estimate this number has obviously to do with the prime factorization of integers and is correspondingly difficult. To simplify it, we group the k_i into levels. Let

$$\ell(k_i) = \max \{ \ell \in \mathbb{Z} \mid 2^\ell \le k_i \}. \tag{8.17}$$

An upper bound for the number of these sequences is then the number of the strictly decreasing finite sequences $k_1 > k_2 > \dots > k_N$ of natural numbers for which

$$\prod_{i=1}^{N} 2^{\ell(k_i)} \le 2^L. \tag{8.18}$$

Since there are at most 2^ℓ numbers k_i for which $\ell(k_i) = \ell$, their number is

$$a(N,L) = \sum_{v} \prod_{\ell=0}^{\infty} \binom{2^\ell}{v(\ell)}, \tag{8.19}$$

where the sum runs over all sequences v of integers $0 \le v(\ell) \le 2^\ell$ for which

$$\sum_{\ell=0}^{\infty} v(\ell) = N, \quad \sum_{\ell=0}^{\infty} v(\ell)\ell \le L. \tag{8.20}$$

The binomial coefficient in (8.19) represents the number of possibilities to choose $v(\ell)$ distinct numbers k_i from the set of the 2^ℓ integers $2^\ell,\dots,2^{\ell+1} - 1$ of level ℓ. To calculate the $a(N,L)$, we introduce the quantities $a(K;N,L)$ that are defined in the same way as the $a(N,L)$ with the exception that the additional condition

$$v(\ell) = 0 \ \text{ for } \ell > K \tag{8.21}$$

is imposed on the sequences v. Since necessarily $v(\ell) = 0$ for $\ell > L$,

$$a(N,L) = a(K;N,L) \ \text{ for } K \ge L. \tag{8.22}$$

The $a(K;N,L)$ can be calculated recursively starting from $a(0;N,L) = 1$ for $N = 0$ and $N = 1$ and $a(0;N,L) = 0$ for all other values of N. For $K \ge 1$,

$$a(K;N,L) = \sum_{v'} \binom{2^K}{v'} a(K-1;N-v',L-Kv'), \tag{8.23}$$

where $v' = v(K)$ runs from 0 to the maximum integer less than or equal 2^K, L/K, and N, the reason being that, assuming (8.21), the conditions (8.20) transfer to

$$\sum_{\ell=0}^{K-1} v(\ell) = N - v(K), \quad \sum_{\ell=0}^{K-1} v(\ell)\ell \le L - Kv(K). \tag{8.24}$$

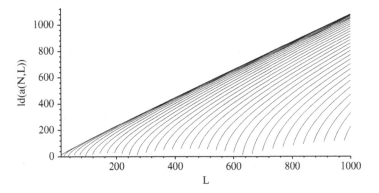

Fig. 8.1 The growth of the numbers $a(N,L)$ for $N = 10, 15, 20, \ldots, 175$

If one steps down from the considered maximum values for L and N to zero, the known quantities $a(K-1;N,L)$ can be directly overwritten with the new values $a(K;N,L)$. The algorithm can thus be easily realized without recursive function calls within every software system allowing for integers of arbitrary length. For

$$N = \sum_{\ell=0}^{m} 2^{\ell} + j, \quad 0 \leq j < 2^{m+1},$$ (8.25)

the minimum L for which there is a sequence v satisfying the conditions (8.20) and with that $a(N,L) > 0$ is that for which the single levels ℓ are maximally filled, that is,

$$L = \sum_{\ell=0}^{m} 2^{\ell} \ell + j(m+1).$$ (8.26)

Let $L(N)$ denote this minimum L assigned to the number N given by (8.25). The $L(N)$ increase very rapidly; for $N = 179$ already $L(N) > 1000$. A crude estimate yields $N \leq L+1$ if $a(N,L) > 0$, or conversely $a(N,L) = 0$ if $L < N-1$. Thus

$$a^*(L) := \max_{N \geq 1} a(N,L) = \max_{N \leq L+1} a(N,L).$$ (8.27)

Figure 8.1 shows, in logarithmic scale, how the $a(N,L)$, extended to piecewise linear functions, behave compared to their joint least upper bound $a^*(L)$. It becomes obvious from this picture that this common upper bound exceeds the actual dimensions for larger N by many orders of magnitude, the more the more N increases.

8.3 An Asymptotic Bound

The best possible upper bound that is independent of N for number of the sequences $k_1 > \ldots > k_N$ of natural numbers k_i for which (8.16) holds grows at least like $\sim 2^L$. The reason for that is that already in the case $N = 1$, there are 2^L such "sequences", those with values $k_1 = 1, \ldots, 2^L$. Figure 8.1 suggests conversely that the upper bound (8.27) for the quantities (8.19), and with that for the given number of the sequences $k_1 > \ldots > k_N$, does not grow much faster than $\sim 2^L$. This is in fact the case as already demonstrated by a rather crude estimate that can be deduced from the following lemma and a well-known result from combinatorics that has its roots in considerations of Euler [26] and was first proved by Hardy and Ramanujan [41].

Lemma 8.3. *The number of the infinite, monotonely decreasing sequences*

$$k_1 \geq k_2 \geq k_3 \geq \ldots \tag{8.28}$$

of natural numbers for which

$$\prod_{i=1}^{\infty} 2^{\ell(k_i)} \leq 2^L, \tag{8.29}$$

with L a given nonnegative integer, is bounded by the quantity

$$\sum_{\ell=0}^{L} p(\ell) 2^{\ell}, \tag{8.30}$$

where $p(\ell)$ denotes the partition number of ℓ, the number of possibilities of representing ℓ as sum of nonnegative integers without regard to the order.

Proof. The number of these sequences is bounded from above by the number of sequences $k_1, k_2, k_3, \ldots$ of natural numbers for which at least their levels $\ell(k_i)$ decrease monotonely and that satisfy (8.29). We show that the expression (8.30) counts the number of these sequences. Let the integers $\ell_i = \ell(k_i)$ first be given. As there are 2^{ℓ_i} natural numbers k_i for which $\ell(k_i) = \ell_i$, namely $k_i = 2^{\ell_i}, \ldots, 2^{\ell_i+1} - 1$, there are

$$\prod_{i=1}^{\infty} 2^{\ell_i} = 2^{\ell}, \quad \ell = \sum_{i=1}^{\infty} \ell_i,$$

sequences $k_1, k_2, k_3, \ldots$ for which the $\ell(k_i)$ attain the prescribed values ℓ_i. The problem therefore reduces to the question how many monotonely decreasing sequences of nonnegative integers ℓ_i exist that sum up to values $\ell \leq L$, that is, for which

$$\sum_{\ell=1}^{\infty} \ell_i = \ell$$

for an $\ell \leq L$. This number is the partition number $p(\ell)$ of ℓ. $\square$

The partition number plays a big role in combinatorics and has first been studied by Euler [26]. Hardy and Ramanujan [41] determined the asymptotic behavior of $p(n)$ as n goes to infinity. One of the simpler estimates they proved reads as follows:

Theorem 8.4. *There is a constant K independent of n such that*

$$p(n) \leq \frac{K}{n} e^{2\sqrt{2n}} \tag{8.31}$$

holds for all natural numbers n.

As the partition numbers $p(n)$ increase monotonely in n and therefore $p(\ell) \leq p(L)$ for all natural numbers $\ell \leq L$, we can conclude from the estimate (8.31) that

$$\lim_{L \to \infty} 2^{-(1+\vartheta)L} \sum_{\ell=0}^{L} p(\ell) 2^\ell = 0 \tag{8.32}$$

for all $\vartheta > 0$. Every sequence $k_1 > k_2 > \ldots > k_N$ of natural numbers for which

$$\prod_{i=1}^{N} k_i \leq 2^L, \tag{8.33}$$

holds can obviously be expanded to an infinite, monotonely decreasing sequence (8.28) that satisfies the condition (8.29) by setting all $k_i = 1$ for $i > N$. The sum (8.30) represents therefore also an upper bound for the number of these sequences. Hence the number of these sequences does indeed not grow faster than

$$\lesssim (2^L)^{1+\vartheta}, \quad \vartheta > 0 \text{ arbitrarily small}, \tag{8.34}$$

independent of N, a value that cannot be substantially improved. The upper bound for the number of these sequences from Sect. 8.2 behaves, because of (8.31), like

$$a^*(L) = (2^L)^{1+\varepsilon(L)}, \quad \varepsilon(L) \leq cL^{-1/2}. \tag{8.35}$$

The exponent $1 + \varepsilon(L)$ can be computed as described there and decays for L ranging from 10 to 1000 monotonely from 1.406 to 1.079. For $L = 100$, $1 + \varepsilon(L) = 1.204$.

8.4   A Proof of the Estimate for the Partition Numbers

The estimate from Theorem 8.4 is by far not the best possible. In fact, Hardy and Ramanujan proved in [41] that the partition number behaves asymptotically like

$$p(n) = \left(\frac{1}{4\sqrt{3}} + o(1) \right) \frac{\exp\left(\pi \sqrt{2n/3}\right)}{n} \tag{8.36}$$

as n goes to infinity. This result has later been improved by Rademacher [67], who has shown that (8.36) is the first term in an infinite series that represents $p(n)$ exactly. We restrict ourselves here to the much simpler proof of the estimate (8.31) that suffices for our purposes and follow hereby the lines given in [41].

Hardy and Ramanujan start from three identities that go back to Euler. Euler first observed that the $p(n)$ are the expansion coefficients of the infinite product

$$\sum_{n=0}^{\infty} p(n) z^n = \prod_{i=1}^{\infty} \frac{1}{1 - z^i}, \tag{8.37}$$

or that this infinite product is in today's terminology their generating function. The partial products of this infinite product converge uniformly on the discs $|z| \le R$ of all radii $R < 1$. The limit function is thus an analytic function that possesses a power series expansion converging for $|z| < 1$. Expanding the single factors as

$$\prod_{i=1}^{\infty} \frac{1}{1 - z^i} = \prod_{i=1}^{\infty} \left(\sum_{k=0}^{\infty} z^{ki} \right), \tag{8.38}$$

one further recognizes that the coefficient in front of z^n is the number of possibilities to represent the number n as a sum $n = k_1 \cdot 1 + \ldots + k_n \cdot n$ of nonnegative integer multiples of $i = 1, \ldots, n$, which is the partition number $p(n)$ of n. This proves (8.37). The difficulties in describing the asymptotic behavior of $p(n)$ for n tending to infinity have a lot to do with the complicated behavior of the infinite product (8.37) when approaching the boundary of the unit circle.

Let $p(n; r)$ denote the number of possibilities to write the nonnegative integer n as an infinite sum $n = n_1 + n_2 + \ldots$ of nonnegative integers $n_1 \ge n_2 \ge \ldots$ with $n_i = 0$ for all indices i greater than r. In the same way one sees then that

$$\sum_{n=0}^{\infty} p(n; r) z^n = \prod_{i=1}^{r} \frac{1}{1 - z^i} \tag{8.39}$$

is the generating function of these restricted partition numbers $p(n; r)$ that play an important role in our argumentation too.

The third identity is a little bit more tricky. Its proof is based on an elementary but ingenious argument from combinatorics. We refer to [5] for such techniques.

Lemma 8.4. *For all complex numbers* $|z| < 1$,

$$\prod_{i=1}^{\infty} \frac{1}{1 - z^i} = 1 + \sum_{r=1}^{\infty} z^{r^2} \prod_{i=1}^{r} \left(\frac{1}{1 - z^i} \right)^2. \tag{8.40}$$

Proof. The proof is based on a classification of the partitions of natural numbers. The idea is to assign to each finite partition $n = n_1 + n_2 + \ldots$, $n_1 \ge n_2 \ge \ldots \ge 1$, a so-called Ferrers diagram that consists of n dots which are arranged in rows and

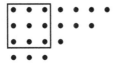

Fig. 8.2 The Ferrers diagram of a partition of the number 20 and the associated Durfee square

columns. The first row consists of n_1 dots, the second of n_2 dots, and so on. The Durfee square of a partition is the largest $r \times r$ square that can be drawn into the upper left corner of its Ferrers diagram. Figure 8.2 shows the Ferrers diagram and the Durfee square assigned to the partition $7+6+4+3$ of the number 20. The partitions of n are classified by the size $r \times r$ of their Durfee squares. Assume that, additionally to the r^2 dots in the Durfee square of a given partition of the number n, there are $k+\ell = n - r^2$ dots in the upper right and the lower left corner of the corresponding Ferrers diagram, k in the upper right and ℓ in the lower left. The upper right corner has at most r rows and corresponds therefore, for $k \geq 1$, to a partition of k into a sum of at most r natural numbers. The number of such partitions is $p(k;r)$. The lower left corner corresponds, for $\ell \geq 1$, to a partition of ℓ whose Ferrers diagram has at most r columns. The number of these partitions is $p(\ell;r)$. The reason for that is that one can assign to every partition of ℓ a conjugate partition interchanging the rows and columns of the associated Ferrers diagram. Since $p(0;r) = 1$, the total number of partitions of a natural number n is thus

$$p(n) = \sum_{1 \leq r^2 \leq n} \sum_{k+\ell=n-r^2} p(k;r)\, p(\ell;r), \tag{8.41}$$

including those with no dot to the right or below the assigned Durfee square. The outer sum classifies the partitions of n by the size of their Durfee squares, and the inner sum is the number of partitions of n with an $r \times r$ Durfee square.

The proof of (8.40) is based on this identity. If one inserts it into the left hand side, splits z^n according to the partition $n = r^2 + k + \ell$, and rearranges the sums, one gets

$$\sum_{n=1}^{\infty} p(n) z^n = \sum_{r=1}^{\infty} z^{r^2} \left(\sum_{k=0}^{\infty} p(k;r) z^k \right) \left(\sum_{\ell=0}^{\infty} p(\ell;r) z^\ell \right).$$

If one inserts the generating function (8.39) for the inner sums on the right hand side and adds the value 1 to both sides, the proposition follows from (8.37). □

Equation (8.40) can be translated into a set of formulas that can be used to calculate the partition numbers, but also form the basis for our subsequent estimates.

Lemma 8.5. *Let $q_1(n) = n+1$ and define $q_r(n)$ for $n \geq 0$ and $r > 1$ recursively by*

$$q_{r+1}(n) = \sum_{(r+1)\ell \leq n} (\ell+1)\, q_r\big(n - (r+1)\ell\big). \tag{8.42}$$

Then the partition numbers $p(n)$ for all natural numbers n are given by

$$p(n) = \sum_{1 \leq r^2 \leq n} q_r(n - r^2). \tag{8.43}$$

Proof. The $q_r(n)$ are the expansion coefficients of the product

$$\prod_{i=1}^{r} \left(\frac{1}{1 - z^i}\right)^2 = \sum_{n=0}^{\infty} q_r(n) z^n,$$

as can be shown by induction on r. If one inserts this relation and (8.37) into (8.40), the relation (8.43) follows by equating the expansion coefficients.   □

The following estimate, from which Hardy and Ramanujan derived their bound for the partition numbers, is based on the representation from Lemma 8.5.

Lemma 8.6. *For all natural numbers n,*

$$p(n) \leq \sum_{r=1}^{\infty} \frac{n^{2r-1}}{(2r-1)!\,(r!)^2}. \tag{8.44}$$

Proof. Following Hardy and Ramanujan, we show first that

$$q_r(n) \leq \frac{(n + r^2)^{2r-1}}{(2r-1)!\,(r!)^2}. \tag{8.45}$$

As $q_1(n) = n + 1$, this is true for $r = 1$. If (8.45) holds for r given, (8.42) yields

$$q_{r+1}(n) \leq \sum_{(r+1)\ell \leq n} (\ell + 1) \frac{(n - (r+1)\ell + r^2)^{2r-1}}{(2r-1)!\,(r!)^2}. \tag{8.46}$$

For $a, b \geq 0$ and all integers $m \geq 2$,

$$m(m-1)a^{m-2}b^2 \leq (a+b)^m - 2a^m + (a-b)^m.$$

Inserting the values $m = 2r + 1$, $a = n - (r+1)\ell + r^2$, and $b = r + 1$ and utilizing the abbreviation $\alpha(\ell) = (n - (r+1)\ell + r^2)^{2r+1}$, one obtains from this inequality

$$(2r+1)2r(r+1)^2 (n - (r+1)\ell + r^2)^{2r-1} \leq \alpha(\ell-1) - 2\alpha(\ell) + \alpha(\ell+1).$$

If we denote by L the maximum integer ℓ for which $(r+1)\ell \leq n$, (8.46) yields

$$(2r+1)!\,((r+1)!)^2 \, q_{r+1}(n) \leq \sum_{\ell=0}^{L} (\ell+1)\big(\alpha(\ell-1) - 2\alpha(\ell) + \alpha(\ell+1)\big)$$

and, evaluating the sum on the right hand side,

$$(2r+1)!\,((r+1)!)^2\,q_{r+1}(n) \leq \alpha(-1)+(L+1)\,\alpha(L+1)-(L+2)\,\alpha(L).$$

If $n-(r+1)(L+1)+r^2 \geq 0$, one has $0 \leq \alpha(L+1) \leq \alpha(L)$. Otherwise $\alpha(L+1) < 0$. Since $\alpha(L) \geq 0$ by the definition of L, one obtains in each of the two cases

$$(2r+1)!\,((r+1)!)^2\,q_{r+1}(n) \leq \alpha(-1) \leq (n+(r+1)^2)^{2r+1},$$

which finishes the proof of (8.45). From (8.43) and (8.45) we get

$$p(n) \leq \sum_{1 \leq r^2 \leq n} \frac{n^{2r-1}}{(2r-1)!\,(r!)^2},$$

which proves the asserted estimate (8.44) for the partition numbers. $\qquad\square$

The rest follows from Stirling's formula that relates factorials to powers and is proven in many introductory analysis textbooks; see for instance [52]. It reads

$$\lim_{n \to \infty} \frac{n^{n+1/2}}{n!\,e^n} = \frac{1}{\sqrt{2\pi}}. \tag{8.47}$$

Stirling's formula yields in the limit of r tending to infinity

$$\lim_{r \to \infty} \frac{1}{(2r-1)!\,(r!)^2} \frac{(4r)!}{2^{6r}} = \frac{\sqrt{2}}{\pi}. \tag{8.48}$$

Hence there exists, by Lemma 8.6, a constant K such that

$$p(n) \leq K \sum_{r=1}^{\infty} \frac{2^{6r}n^{2r-1}}{(4r)!} = \frac{K}{n} \sum_{r=1}^{\infty} \frac{(2\sqrt{2n})^{4r}}{(4r)!}. \tag{8.49}$$

The proposition, that is, the estimate (8.31) from Theorem 8.4 follows from the power series expansion of the exponential function.

8.5 The Complexity of the Quantum N-Body Problem

Our estimates demonstrate that, for the case that all electrons have the same spin, the number of antisymmetrized tensor products or Slater determinants built from the three-dimensional eigenfunctions that are needed to reach an H^1-error of order $\mathcal{O}(\varepsilon)$ does not increase much faster than $\mathcal{O}(\varepsilon^{-6/\alpha})$ for ε tending to zero, where, of course, nothing is said about the constant and its dependence on the different problem parameters. That is, the rate of convergence expressed in terms of the number of basis functions astonishingly does not deteriorate with the space dimension $3N$ or the number N of electrons. It behaves almost as with the expansion of a one-electron wave function into eigenfunctions of the given type. In the case of the expansion into Gaussians, the eigenfunctions of the harmonic oscillator (8.3), the constant α attains the value $\alpha = 1$, a value that we will still improve to $\alpha = 3/2$. The results from Sect. 8.1 and particularly Theorem 8.3 show that one can come arbitrarily close to $\alpha = 2$, but cannot completely reach this value. The rate with which the dimension of the corresponding spaces grows with increasing accuracy then behaves asymptotically almost like that of a first-order method in three space dimensions. In the general case of electrons of distinct spin, the order of convergence halves due to the singularities of the wave functions at the places where electrons with opposite spin meet, which is reflected in the presence of two products instead of only one in (7.28) and (7.30) or the square root in (7.31). The rate of convergence remains, however, independent of the number of electrons and comes arbitrarily close to that for the two-electron case. Our considerations thus show that the complexity of the quantum-mechanical N-body problem is much lower than generally believed.

Chapter 9
The Radial-Angular Decomposition

Symmetry plays an important role in quantum mechanics. Closed solutions of quantum mechanical problems are mostly determined with help of symmetry properties of the underlying Schrödinger equation. This holds particularly for one-electron problems with rotationally symmetric potentials. In this case the solutions split into products of problem-dependent radial parts and angular parts that are built up from three-dimensional spherical harmonics. The most prominent example is the Schrödinger equation for hydrogen-like atoms. The knowledge about its solutions is basic for our understanding of chemistry. The solutions of the Schrödinger equation for a general system of N electrons moving in the field of a given number of clamped nuclei unfortunately do not attain such a simple form. The norms that we introduced to measure their mixed derivatives are however invariant to rotations of the coordinates of the single electrons. We therefore decompose the solutions of the N-particle equation in this chapter into tensor products of three-dimensional angular momentum eigenfunctions, the decomposition that reflects this rotational invariance. The contributions of these tensor products to the total energy decrease like

$$\left\{ \prod_{i \in I_-} \left(1 + \ell_i(\ell_i + 1) \right) + \prod_{i \in I_+} \left(1 + \ell_i(\ell_i + 1) \right) \right\}^{-1} \tag{9.1}$$

with the angular momentum quantum numbers ℓ_i of the electrons; I_- and I_+ are again the sets of the indices of the electrons with spin $\pm 1/2$. We will use this decomposition to study the convergence behavior of the eigenfunction expansions from the last chapters further and in particular will obtain an improved estimate for the convergence rate of the expansion into Gauss functions, the eigenfunctions of the harmonic oscillator. The central sections of this chapter, in which N-particle wave functions are studied, are Sect. 9.2, Sect. 9.6, and Sect. 9.7. The considerations there are based on the examination of the three-dimensional case, to which most of this chapter is devoted. In the first section the decomposition into the eigenfunctions of the three-dimensional angular momentum operator, the spherical harmonics is studied. The third section treats the three-dimensional radial Schrödinger equation in general, and the following two the Coulomb problem and the harmonic oscillator.

H. Yserentant, *Regularity and Approximability of Electronic Wave Functions*,
Lecture Notes in Mathematics 2000, DOI 10.1007/978-3-642-12248-4_9,
© Springer-Verlag Berlin Heidelberg 2010

9.1   Three-Dimensional Spherical Harmonics

Quantum mechanical operators result from their classical counterpart via the correspondence principle. The classical angular momentum $L = x \times p$, p the momentum,
is a vector-valued quantity, or in three space dimensions can at least be interpreted as
such a quantity. Since the quantum mechanical momentum operator is $p = -i\hbar\nabla$,
the quantum mechanical angular momentum operator is therefore the vector valued
operator $L = -i\hbar x \times \nabla$. Its square reads in atomic units, in which $\hbar = 1$,

$$L^2 = -\frac{1}{2} \sum_{\substack{i,j=1 \\ i \neq j}}^{3} \left(x_i \frac{\partial}{\partial x_j} - x_j \frac{\partial}{\partial x_i} \right)^2, \tag{9.2}$$

where x_1, x_2, and x_3 denote in this section the components of $x \in \mathbb{R}^3$. The aim of this
section is to decompose functions from $\mathbb{R}^3$ to $\mathbb{R}$ into eigenfunctions of this operator
and to study the convergence of the corresponding expansions in Sobolev spaces of
arbitrary order. These eigenfunctions split into products of rotationally symmetric
functions and spherical harmonics, functions that cover their angular dependence.

Lemma 9.1. *The operator* L^2 *is formally self-adjoint, in the sense that for all infinitely differentiable functions* $u, v : \mathbb{R}^3 \to \mathbb{R}$ *with compact support*

$$(L^2 u, v) = (u, L^2 v). \tag{9.3}$$

Proof. Integration by parts yields for all indices $i \neq j$

$$\int (x_i D_j u - x_j D_i u)\, v\, dx = -\int u\,(x_i D_j v - x_j D_i v)\, dx$$

so that already the single terms of which the sum in the definition (9.2) of the operator is composed have the asserted property.   □

The operator L^2 does not act on the radial part of a function and leaves it untouched:

Lemma 9.2. *If* $f : \mathbb{R}_{>0} \to \mathbb{R}$ *and* $\phi : \mathbb{R}^3 \to \mathbb{R}$ *are infinitely differentiable and*

$$u(x) = f(r)\phi(x), \quad r = |x|, \tag{9.4}$$

the function $L^2 u$ *is given by*

$$L^2 u = f L^2 \phi. \tag{9.5}$$

Proof. Since the two terms involving the derivative of f cancel,

$$\left(x_i \frac{\partial}{\partial x_j} - x_j \frac{\partial}{\partial x_i} \right) \{ f(r)\phi(x) \} = f(r)\left(x_i \frac{\partial}{\partial x_j} - x_j \frac{\partial}{\partial x_i} \right)\phi(x),$$

from which the proposition follows.   □

Lemma 9.3. *If $f : \mathbb{R}_{>0} \to \mathbb{R}$ is an infinitely differentiable function and H_ℓ a harmonic polynomial that is homogeneous of degree ℓ, the function given by*

$$u(x) = f(r) H_\ell(x), \quad r = |x|, \tag{9.6}$$

is an eigenfunction of the operator L^2 for the eigenvalue $\ell(\ell+1)$ in the sense that

$$\mathrm{L}^2 u = \ell(\ell+1) u \tag{9.7}$$

holds on the domain of definition of u.

Proof. The proof is based on the representation

$$\mathrm{L}^2 = -r^2 \Delta + \sum_{i,j=1}^{3} x_i x_j D_i D_j + 2 \sum_{i=1}^{3} x_i D_i$$

that can be verified with help of the elementary differentiation rules. Moreover,

$$\frac{d}{d\lambda} H_\ell(\lambda x) = \sum_{i=1}^{3} x_i (D_i H_\ell)(\lambda x), \quad \frac{d^2}{d\lambda^2} H_\ell(\lambda x) = \sum_{i,j=1}^{3} x_i x_j (D_i D_j H_\ell)(\lambda x).$$

Since H_ℓ is homogeneous of degree ℓ, that is, $H_\ell(\lambda x) = \lambda^\ell H_\ell(x)$, on the other hand

$$\frac{d}{d\lambda} H_\ell(\lambda x) = \ell \lambda^{\ell-1} H_\ell(x), \quad \frac{d^2}{d\lambda^2} H_\ell(\lambda x) = (\ell-1)\ell \lambda^{\ell-2} H_\ell(x).$$

Setting $\lambda = 1$ we obtain the relations

$$\sum_{i=1}^{3} x_i D_i H_\ell = \ell H_\ell, \quad \sum_{i,j=1}^{3} x_i x_j D_i D_j H_\ell = (\ell-1)\ell H_\ell$$

that express the homogeneity of the polynomial H_ℓ. As by assumption $\Delta H_\ell = 0$,

$$\mathrm{L}^2 H_\ell = \ell(\ell+1) H_\ell$$

follows. With that the proposition results from Lemma 9.2. □

The function given by (9.6) can be written in the form

$$u(x) = r^\ell f(r) K_\ell(x), \quad r = |x|, \tag{9.8}$$

i.e., splitting into the rotationally symmetric radial part $x \to r^\ell f(r)$ and the function

$$K_\ell(x) = H_\ell\left(\frac{x}{r}\right) \tag{9.9}$$

that covers its angular dependence. The functions of the form (9.9) are the spherical harmonics of degree ℓ. They are homogeneous of degree 0 and satisfy the equation

$$L^2 K_\ell = \ell(\ell+1) K_\ell, \tag{9.10}$$

that is, are in this sense themselves eigenfunctions of the operator (9.2). The symmetry of this operator implies the following first orthogonality property:

Lemma 9.4. *If $f,g : \mathbb{R}^3 \to \mathbb{R}$ are rotationally symmetric, infinitely differentiable functions with compact support and H_ℓ and $H_{\ell'}$ harmonic polynomials of degrees ℓ and $\ell' \neq \ell$, the functions*

$$u(x) = f(x) H_\ell(x), \quad v(x) = g(x) H_{\ell'}(x) \tag{9.11}$$

are orthogonal to each other with respect to the L_2-inner product.

Proof. By the just proven Lemma 9.3

$$L^2 u = \ell(\ell+1) u, \quad L^2 v = \ell'(\ell'+1) v$$

and therefore, by Lemma 9.1,

$$\ell(\ell+1)(u,v) = (L^2 u, v) = (u, L^2 v) = \ell'(\ell'+1)(u,v).$$

As $\ell'(\ell'+1) \neq \ell(\ell+1)$ for $\ell' \neq \ell$, the proposition $(u,v) = 0$ follows. $\square$

To proceed, we need to integrate corresponding functions $x \to f(x_1, x_2, x_3)$ over the unit sphere S consisting of the vectors $x \in \mathbb{R}^3$ of length $|x| = 1$. Such integrals are of course invariant to rotations. Their parameter representation in polar coordinates is

$$\int_S f \, dx = \int_0^{2\pi} \int_{-\pi/2}^{\pi/2} f(\cos \varphi \cos \vartheta, \sin \varphi \cos \vartheta, \sin \vartheta) \cos \vartheta \, d\vartheta \, d\varphi. \tag{9.12}$$

The polar coordinate representation of integrals over the $\mathbb{R}^3$ reads in this notation

$$\int_{\mathbb{R}^3} u(x) \, dx = \int_0^\infty \left\{ r^2 \int_S u(r\eta) \, d\eta \right\} dr, \tag{9.13}$$

where the inner integral is the integral of the function $\eta \to u(r\eta)$ depending on the parameter r over the unit sphere S and can be further resolved into a double integral.

Lemma 9.5. *Let K_ℓ and $K_{\ell'}$ be spherical harmonics of degrees ℓ and $\ell' \neq \ell$. Then*

$$\int_S K_\ell(\eta) K_{\ell'}(\eta) \, d\eta = 0. \tag{9.14}$$

Proof. Let $f : \mathbb{R}^3 \to \mathbb{R}$ be an infinitely differentiable, rotationally symmetric function with compact support and let H_ℓ and $H_{\ell'}$ be the harmonic polynomials that are assigned to the spherical harmonics. Let n be arbitrarily given fixed unit vector. Then

$$\int f(x) H_\ell(x) f(x) H_{\ell'}(x) \, dx = \left(\int_0^\infty r^{2+\ell+\ell'} f(rn)^2 \, dr \right) \left(\int_S H_\ell(\eta) H_{\ell'}(\eta) \, d\eta \right),$$

as follows from (9.13) and the properties of the functions f, H_ℓ, and $H_{\ell'}$. Since the integral on the left hand side vanishes by Lemma 9.4 and the first integral on the right hand side takes a value greater than zero as long as f is different from zero, the second integral on the right hand side must vanish.   □

Lemma 9.6. *The space of the homogeneous harmonic polynomials of degree ℓ, and with that also the assigned space of spherical harmonics, has the dimension $2\ell + 1$.*

Proof. For $\ell = 0$ and $\ell = 1$, every homogeneous polynomial is harmonic and the corresponding spaces have the asserted dimensions $2\ell + 1 = 1$ and $2\ell + 1 = 3$. For $\ell \geq 2$, we utilize that every such polynomial can be written in the form

$$H_\ell(x) = \sum_{j=0}^{\ell} x_3^j P_{\ell-j}(x_1, x_2),$$

with the $P_{\ell-j}$ polynomials in x_1 and x_2 that are homogeneous of degree $\ell - j$. Thus

$$\Delta H_\ell = \sum_{j=0}^{\ell-2} x_3^j \left(\Delta_2 P_{\ell-j} + (j+1)(j+2) P_{\ell-j-2} \right),$$

where Δ_2 denotes the two-dimensional Laplace operator acting upon the components x_1 and x_2. The requirement $\Delta H_\ell = 0$ is therefore equivalent to the condition

$$P_{\ell-j-2} = \frac{1}{(j+1)(j+2)} \Delta P_{\ell-j}$$

for $j = 0, \ldots, \ell - 2$. That is, P_ℓ and $P_{\ell-1}$ can be arbitrarily given and determine then the other polynomials $P_{\ell-j}$ and with that also H_ℓ. The proposition thus follows from the observation that the space of the polynomials in the variables x_1 and x_2 that are homogeneous of degree n has the dimension $n + 1$.   □

In the sequel we will use an $L_2(S)$-orthonormal basis

$$K_\ell^m(x), \quad m = -\ell, \ldots, \ell, \tag{9.15}$$

of the space of the spherical harmonics of degree ℓ which we do not specify further. Not surprisingly such bases can best be represented in terms of polar coordinates; see the appendix of this chapter. The K_ℓ^m are by definition homogeneous of degree 0 and represent the angular parts of the homogeneous harmonic polynomials

$$H_\ell^m(x) = r^\ell K_\ell^m(x), \quad r = |x|, \tag{9.16}$$

of degree ℓ. The spherical harmonics K_ℓ^m, $\ell = 0,1,2,\ldots$, $m = -\ell,\ldots,\ell$, form an $L_2(S)$-orthonormal basis of the space of all spherical harmonics, as follows from Lemma 9.5, and the assigned polynomials (9.16) correspondingly a basis of the space of all harmonic polynomials in three variables.

Lemma 9.7. *Every polynomial P_ℓ that is homogeneous of degree ℓ can be written as*

$$P_\ell(x) = \sum_{0 \le j \le \ell/2} |x|^{2j} H_{\ell-2j}(x), \qquad (9.17)$$

where the H_{l-2j} are harmonic polynomials of degree $\ell - 2j$.

Proof. As follows from Lemma 9.5, a polynomial of the form (9.17) vanishes if and only if the single terms of which the sum is composed vanish individually. Therefore the space of the polynomials (9.17) has, by Lemma 9.6, the dimension

$$\sum_{0 \le j \le \ell/2} (2(\ell - 2j) + 1) = \frac{(\ell+1)(\ell+2)}{2},$$

which coincides with the dimension of the space of all polynomials in three variables that are homogeneous of degree ℓ, a fact from which the proposition follows. □

We can finally state that the linearly independent polynomials

$$|x|^{2n} H_\ell^m(x) = |x|^{2n+\ell} K_\ell^m(x), \quad n, \ell = 0,1,2,\ldots, \; m = -\ell,\ldots,\ell, \qquad (9.18)$$

span the space of all polynomials in three variables and that every polynomial coincides on the unit sphere with a harmonic polynomial of at most the same degree.

Let $\mathcal{V}_\ell^m$ be the infinite-dimensional space that consists of the functions

$$x \to f(r) H_\ell^m(x), \quad r = |x|, \qquad (9.19)$$

with $f : \mathbb{R} \to \mathbb{R}$ an arbitrary infinitely differentiable function with compact support. The spaces $\mathcal{V}_\ell^m$ are L_2-orthogonal to each other. The L_2-inner product of functions

$$u(x) = f(r) H_\ell^m(x), \quad v(x) = g(r) H_\ell^m(x) \qquad (9.20)$$

in $\mathcal{V}_\ell^m$ can be reduced to the one-dimensional integral

$$\int_{\mathbb{R}^3} u(x) v(x) \, dx = \int_0^\infty r^{2+2\ell} f(r) g(r) \, dr. \qquad (9.21)$$

An immediate consequence of this observation is:

Lemma 9.8. *The closure of $\mathcal{V}_\ell^m$ in L_2 consists of the functions*

$$x \to \frac{1}{r} f(r) K_\ell^m(x), \qquad (9.22)$$

with functions $f : \mathbb{R}_{>0} \to \mathbb{R}$ that are square integrable over the positive real axis.

To decompose functions into radial and angular parts, we assign to every infinitely differentiable function $u : \mathbb{R}^3 \to \mathbb{R}$ with compact support the functions given by

$$(Q_\ell^m u)(x) = \left\{ \int_S u(r\eta) K_\ell^m(\eta) \, d\eta \right\} K_\ell^m(x), \qquad (9.23)$$

where again the abbreviation $r = |x|$ has been used.

Lemma 9.9. *The functions $Q_\ell^m u$ belong themselves to the spaces $\mathscr{V}_\ell^m$ and are the L_2-orthogonal projections of the given functions u onto these.*

Proof. We fix the function u and the indices ℓ and m and study first the radial part

$$f(r) = \int_S u(r\eta) K_\ell^m(\eta) \, d\eta$$

of $Q_\ell^m u$. We claim that, for $\ell \geq 1$, it can be written in the form

$$f(r) = r^\ell g(r)$$

with the infinitely differentiable function

$$g(r) = \frac{1}{(\ell-1)!} \int_0^1 (1-\vartheta)^{\ell-1} f^{(\ell)}(\vartheta r) \, d\vartheta.$$

This follows from the fact that

$$\left(\frac{d}{dr} \right)^j u(r\eta) \Big|_{r=0} = \sum_{|\alpha|=j} (D^\alpha u)(0) \, \eta^\alpha$$

and therefore $f^{(j)}(0) = 0$ for $j = 0, \ldots, \ell-1$ due to the $L_2(S)$-orthogonality of K_ℓ^m to every polynomial of degree less than ℓ. The integral form of Taylor's theorem yields the above representation of f, that is, the desired representation

$$(Q_\ell^m u)(x) = g(r) H_\ell^m(x)$$

of the function $Q_\ell^m u$. The difference $u - Q_\ell^m u$ is L_2-orthogonal to the functions in $\mathscr{V}_\ell^m$, as can be seen representing the L_2-inner product in term of polar coordinates. That is, $Q_\ell^m u$ is indeed the L_2-orthogonal projection of u onto $\mathscr{V}_\ell^m$. $\qquad \square$

As a consequence the operators Q_ℓ^m can be extended to L_2-orthogonal projections from L_2 to the closure of $\mathscr{V}_\ell^m$ in L_2. The key result, into the proof of which our knowledge on the eigenfunctions of the harmonic oscillator enters, is:

Theorem 9.1. *The functions $u \in L_2(\mathbb{R}^3)$ possess the L_2-orthogonal decomposition*

$$u = \sum_{\ell=0}^{\infty} \sum_{m=-\ell}^{\ell} Q_\ell^m u. \qquad (9.24)$$

Proof. In view of Lemma 9.9 it suffices to prove that every square integrable function can be approximated arbitrarily well in the L_2-sense by a finite linear combination of functions in the spaces $\mathscr{V}_\ell^m$. We already know from the discussion in Sect. 3.4 that the finite linear combinations of the eigenfunctions of the three-dimensional harmonic oscillator form a dense subset of L_2. These eigenfunctions span the linear space of the products (3.85) of polynomials with a fixed Gaussian and can therefore be represented as finite linear combinations of the functions

$$x \rightarrow |x|^{2n} e^{-|x|^2/2} H_\ell^m(x).$$

Thus it suffices to show that these can be approximated arbitrarily well by functions in the spaces $\mathscr{V}_\ell^m$. But this is readily seen, simply by multiplying them with a series of infinitely differentiable, rotationally symmetric cut-off functions.   □

The series (9.24) is, for sufficiently smooth functions u, not only an orthogonal decomposition in L_2 but in every Sobolev space of corresponding order. This is based on the fact that the projections Q_ℓ^m and the Laplace operator commute:

Theorem 9.2. *If u is an infinitely differentiable function with compact support, so are its projections $Q_\ell^m u$ onto the spaces $\mathscr{V}_\ell^m$. For all such functions u,*

$$\Delta Q_\ell^m u = Q_\ell^m \Delta u. \tag{9.25}$$

Proof. Let φ be an infinitely differentiable test function with compact support that vanishes on a neighborhood of the origin to avoid problems with the potential singularities there. The projection $Q_\ell^m \varphi$ of φ onto $\mathscr{V}_\ell^m$ can then be written in the form

$$(Q_\ell^m \varphi)(x) = f(r) H_\ell^m(x),$$

with $f : \mathbb{R} \rightarrow \mathbb{R}$ an infinitely differentiable function with compact support that vanishes on a neighborhood of the point $r = 0$. Since $x \cdot \nabla H_\ell^m = \ell H_\ell^m$ and $\Delta H_\ell^m = 0$,

$$(\Delta Q_\ell^m \varphi)(x) = \left(f''(r) + \frac{2\ell + 2}{r} f'(r) \right) H_\ell^m(x).$$

That is, $\Delta Q_\ell^m \varphi$ is contained in $\mathscr{V}_\ell^m$, too. We conclude that

$$(Q_\ell^m \Delta u, \varphi) = (u, \Delta Q_\ell^m \varphi) = (Q_\ell^m u, \Delta Q_\ell^m \varphi) = (\Delta Q_\ell^m u, Q_\ell^m \varphi).$$

By the same calculation as above the function $\Delta Q_\ell^m u$ coincides outside every given neighborhood of the origin with a function in $\mathscr{V}_\ell^m$. Let $\Delta Q_\ell^m u = v$, $v \in \mathscr{V}_\ell^m$, everywhere where the given function φ or its projection $Q_\ell^m \varphi$ take a value $\neq 0$. Then

$$(\Delta Q_\ell^m u, Q_\ell^m \varphi) = (v, Q_\ell^m \varphi) = (v, \varphi) = (\Delta Q_\ell^m u, \varphi).$$

Hence, for all test functions φ vanishing on a neighborhood of the origin,

$$(Q_\ell^m \Delta u, \varphi) = (Q_\ell^m u, \Delta \varphi).$$

The next step is to show that this relation still holds even if φ does not vanish on a neighborhood of the origin. For that purpose let χ be an infinitely differentiable function that takes the values $\chi(x) = 0$ for $|x| \leq 1$ and $\chi(x) = 1$ for $|x| \geq 2$. Set $\chi_\varepsilon(x) = \chi(x/\varepsilon)$ for $\varepsilon > 0$. For all infinitely differentiable functions φ then

$$(Q_\ell^m \Delta u, \chi_\varepsilon \varphi) = (Q_\ell^m u, \Delta(\chi_\varepsilon \varphi)) = (Q_\ell^m u, \chi_\varepsilon \Delta \varphi + 2\nabla \chi_\varepsilon \cdot \nabla \varphi + \varphi \Delta \chi_\varepsilon).$$

As $|\nabla \chi_\varepsilon(x)| \leq c/r$ and $|\Delta \chi_\varepsilon(x)| \leq c/r^2$ with a constant c independent of ε, as the derivatives of χ_ε vanish outside the ball of radius 2ε around the origin, and as $Q_\ell^m \Delta u$ and $Q_\ell^m u$ are bounded functions with bounded support, one obtains the desired result from the dominated convergence theorem letting ε tend to zero.

The relation above can obviously be iterated. For all rapidly decreasing φ

$$(Q_\ell^m \Delta^s u, \varphi) = (Q_\ell^m u, \Delta^s \varphi), \quad s = 1, 2, 3, \ldots.$$

To prove that $Q_\ell^m u$ is infinitely differentiable we switch to the Fourier representation of this relation. Plancherel's theorem yields, because of $F\Delta^s \varphi = (-1)^s |\omega|^{2s} F\varphi$,

$$\int F Q_\ell^m \Delta^s u \, \overline{F\varphi} \, d\omega = \int (-1)^s |\omega|^{2s} F Q_\ell^m u \, \overline{F\varphi} \, d\omega$$

for all rapidly decreasing φ. As every infinitely differentiable function with compact support is Fourier transform of a rapidly decreasing function this means, by Lemma 2.4, that the function $F Q_\ell^m \Delta^s u \in L_2$ and the locally integrable function

$$\omega \rightarrow (-1)^s |\omega|^{2s} (F Q_\ell^m u)(\omega)$$

coincide. The latter is therefore square integrable and $Q_\ell^m u$ thus contained in the Sobolev spaces H^{2s} for all $s \in \mathbb{N}$. From Theorem 2.12 we can therefore conclude that $Q_\ell^m u$ is infinitely differentiable. Integration by parts yields

$$(Q_\ell^m \Delta u, \varphi) = (\Delta Q_\ell^m u, \varphi)$$

for all test functions φ and therefore finally (9.25). □

Theorem 9.2 is the main tool to prove the convergence of the derivatives of arbitrary order of the projections to the corresponding derivatives of the function itself:

Theorem 9.3. *The L_2-orthogonal decomposition (9.24) of an infinitely differentiable function u with compact support is also orthogonal with respect to each of the positive semidefinite inner products*

$$(u, v)_s = (-1)^s (\Delta^s u, v), \quad s = 0, 1, 2, \ldots, \tag{9.26}$$

that induce the seminorms $|\cdot|_s$. *That is, for all such functions u and all such s*

$$|u|_s^2 = \sum_{\ell=0}^{\infty} \sum_{m=-\ell}^{\ell} |Q_\ell^m u|_s^2. \qquad (9.27)$$

Proof. Expanding the second argument one obtains for the given functions u

$$(\Delta^s u, u) = \sum_{\ell=0}^{\infty} \sum_{m=-\ell}^{\ell} (\Delta^s u, Q_\ell^m u).$$

From (9.25), that is, the fact that Δ^s and Q_ℓ^m commute, the relation

$$(\Delta^s u, Q_\ell^m u) = (Q_\ell^m \Delta^s u, Q_\ell^m u) = (\Delta^s Q_\ell^m u, Q_\ell^m u)$$

follows. Inserting this relation into the double sum above the proof is finished. $\square$

The norms on the Sobolev spaces H^s, $s = 0, 1, 2, \ldots$, can be composed of the seminorms above. As the infinitely differentiable functions with compact support are dense in these spaces H^s, the theorem shows that (9.24) is not only an orthogonal decomposition of L_2 but of all these Sobolev spaces.

To end this section, we again bring the operator (9.2) into play and begin with the observation is that, for all indices $i \neq j$,

$$(x_i D_j - x_j D_i) \Delta = \Delta (x_i D_j - x_j D_i), \qquad (9.28)$$

from which the commutation relation

$$L^2 \Delta = \Delta L^2 \qquad (9.29)$$

follows. Together with Theorem 9.2 it yields our third decomposition theorem:

Theorem 9.4. *For all infinitely differentiable functions u with compact support and all nonnegative integers s*

$$(L^2 u, u)_s = \sum_{\ell=0}^{\infty} \sum_{m=-\ell}^{\ell} \ell(\ell+1) |Q_\ell^m u|_s^2. \qquad (9.30)$$

Proof. Expanding the second argument one obtains for the given functions u

$$(\Delta^s L^2 u, u) = \sum_{\ell=0}^{\infty} \sum_{m=-\ell}^{\ell} (\Delta^s L^2 u, Q_\ell^m u).$$

From the commutation relation (9.29), Lemma 9.1, and Lemma 9.3

$$(\Delta^s L^2 u, Q_\ell^m u) = (L^2 \Delta^s u, Q_\ell^m u) = (\Delta^s u, L^2 Q_\ell^m u) = \ell(\ell+1)(\Delta^s u, Q_\ell^m u)$$

follows. Since by (9.25), that is, the fact that Δ^s and Q_ℓ^m commute, again

$$(\Delta^s u, Q_\ell^m u) = (Q_\ell^m \Delta^s u, Q_\ell^m u) = (\Delta^s Q_\ell^m u, Q_\ell^m u),$$

which proves the proposition.   □

Finally we observe that, for infinitely differentiable u with bounded support,

$$(L^2 u, u)_s = \frac{1}{2} \sum_{\substack{i,j=1 \\ i \neq j}}^{3} |(x_i D_j - x_j D_i)u|_s^2, \tag{9.31}$$

which follows from (9.28), the definition of the operator L^2, and the skew-symmetry of the operators $x_i D_j - x_j D_i$ that already entered into the proof of Lemma 9.1.

9.2 The Decomposition of N-Particle Wave Functions

The goal of this section is to expand N-particle wave functions, that is, functions

$$u : (\mathbb{R}^3)^N \to \mathbb{R} : (x_1, \ldots, x_N) \to u(x_1, \ldots, x_N) \tag{9.32}$$

depending on the positions $x_i \in \mathbb{R}^3$ of the single electrons, into series of functions

$$x \to f(r_1, \ldots, r_N) K_{\ell_1}^{m_1}(x_i) \ldots K_{\ell_N}^{m_N}(x_i), \quad r_i = |x_i|, \tag{9.33}$$

and to study the convergence properties of these series. The expansion coefficients $f : \mathbb{R}^N \to \mathbb{R}$ can in this case be obtained in the same way as in the previous section applying the operators (9.23) particle-wise. Let $Q(\ell, m)$ be, for multi-indices ℓ and m in $\mathbb{Z}^N$ with components $\ell_i \geq 0$ and $|m_i| \leq \ell_i$, the corresponding projection operator that maps the infinitely differentiable functions (9.32) into the space $\mathscr{V}(\ell, m)$ of the functions (9.33) with infinitely differentiable radial parts $f : \mathbb{R}^N \to \mathbb{R}$. The results from the previous section transfer then more or less immediately to the present situation. The operators $Q(\ell, m)$ can be extended to L_2-orthogonal projections onto the L_2-closure of $\mathscr{V}(\ell, m)$. Since the spaces $\mathscr{V}(\ell, m)$ are orthogonal to each other and every square integrable function can be approximated arbitrarily well in the L_2-sense by a finite linear combination of the functions in these spaces, one obtains:

Theorem 9.5. *The functions $u \in L_2(\mathbb{R}^{3N})$ possess the L_2-orthogonal decomposition*

$$u = \sum_\ell \sum_m Q(\ell, m)u, \tag{9.34}$$

where the outer sum runs over the multi-indices $\ell \in \mathbb{Z}^N$ with components $\ell_i \geq 0$ and the inner sum over the multi-indices $m \in \mathbb{Z}^N$ with components $|m_i| \leq \ell_i$.

The second important result generalizes Theorem 9.2 and will us allow to estimate higher-order norms of the projections $Q(\ell,m)u$ of sufficiently smooth functions u:

Theorem 9.6. *The operators $Q(\ell,m)$ map the space $\mathscr{D}$ of the infinitely differentiable functions with compact support into itself. They commute on this space with every differential operator $\Delta_1^{\alpha_1}\ldots\Delta_N^{\alpha_N}$ with arbitrary nonnegative integer exponents α_i.*

Proof. The key to the proof is the observation that

$$(Q(\ell,m)u,\Delta_i\varphi) = (Q(\ell,m)\Delta_i u,\varphi)$$

holds for all $u \in \mathscr{D}$ and all rapidly decreasing functions φ. This results from the corresponding relation in three space dimensions, on which the proof of Theorem 9.2 is based, with help of Fubini's theorem. Adding these equations up one gets

$$(Q(\ell,m)u,\Delta\varphi) = (Q(\ell,m)\Delta u,\varphi).$$

From here one can proceed as in the proof of Theorem 9.2 and show with help of Theorem 2.12 that the projections $Q(\ell,m)u$ are infinitely differentiable. Hence

$$\Delta_i Q(\ell,m)u = Q(\ell,m)\Delta_i u$$

for all electron indices i, from which the rest follows by induction. $\square$

The counterpart of Theorem 9.3, whose proof is analogously to the proof of this theorem directly based on these properties of the projection operators, is:

Theorem 9.7. *The L_2-orthogonal decomposition (9.34) of a function $u \in \mathscr{D}$ is also orthogonal with respect to every positive semidefinite inner product of the form*

$$\langle u,v \rangle = (-1)^{\alpha_1+\ldots+\alpha_N}(\Delta^{\alpha_1}\ldots\Delta^{\alpha_N}u,v), \tag{9.35}$$

with integer exponents $\alpha_i \geq 0$. The induced seminorm of u splits into the sum

$$|u|^2 = \sum_\ell \sum_m |Q(\ell,m)u|^2. \tag{9.36}$$

This property is inherited by every seminorm or norm that is composed of such seminorms, and by the functions in the completions of $\mathscr{D}$ under such norms. Examples are the H^1-norm, and the norms (6.79) and (6.80) that we introduced in Sect. 6.5 to measure the regularity of the solutions of the electronic Schrödinger equation. The regularity of these solutions thus transfers to their projections. Moreover, since the exponential weight functions from Sect. 6.1 that we introduced to measure the decay of the mixed derivatives split into a product of factors that are invariant under rotations of the electron positions x_i, the projections of the exponentially weighted solutions are weighted projections for the same weight function. The mixed derivatives of their projections show therefore the same kind of decay behavior as the corresponding derivatives of the solutions themselves.

The projections $Q(\ell, m)u$ are eigenfunctions of the operators L_i^2 that are the counterparts of the operator (9.2) acting upon the components of the position vector $x_i \in \mathbb{R}^3$ of the electron i. We assign the differential operator

$$\mathscr{L} = \prod_{i \in I_-} \left(1 + L_i^2\right) + \prod_{i \in I_+} \left(1 + L_i^2\right) \tag{9.37}$$

to the given sets I_- of the indices of the electrons with spin $-1/2$ and I_+ of the indices of the electrons with spin $+1/2$. The operators Δ_i commute, by (9.29), with the operators L_i^2 and thus also with $\mathscr{L}$. The formal self-adjointness of the single parts L_i^2 and with that also of $\mathscr{L}$ leads therefore, as in the proof of Theorem 9.4, to:

Theorem 9.8. *For all infinitely differentiable functions u with compact support*

$$\langle \mathscr{L}u, u \rangle = \sum_\ell \sum_m \left\{ \prod_{i \in I_-} \left(1 + \ell_i(\ell_i + 1)\right) + \prod_{i \in I_+} \left(1 + \ell_i(\ell_i + 1)\right) \right\} |Q(\ell, m)u|^2, \tag{9.38}$$

where the brackets on the left hand side denote any of the inner products (9.35) and the seminorm or norm on the right hand side is induced by this inner product.

This result transfers again to every inner product that is composed of parts of the given kind and in particular to the L_2- and the H^1-inner product to which we restrict ourselves in the sequel. The first order differential operators of which the operators L_i^2 and with that also the operator (9.37) are composed can, as in the proof of (9.31), be distributed to both sides of the inner product. The inner product on the left hand side can thus be estimated in the given case by the L_2-norms of polynomial multiples of the mixed derivatives considered in Theorem 6.13. This, and the fact that the infinitely differentiable functions of corresponding symmetry having a compact support are dense in the spaces in which the solutions are contained, prove:

Theorem 9.9. *Let u be a solution of the electronic Schrödinger equation in the Hilbert space $H^1(\sigma)$ assigned to the given spin distribution for an eigenvalue below the ionization threshold. Then its norm given by the expression*

$$|||u|||^2 = \sum_\ell \sum_m \left\{ \prod_{i \in I_-} \left(1 + \ell_i(\ell_i + 1)\right) + \prod_{i \in I_+} \left(1 + \ell_i(\ell_i + 1)\right) \right\} \|Q(\ell, m)u\|_1^2 \tag{9.39}$$

remains finite, where the size of this norm depends on the degree of excitation.

This is one of our central results and another important consequence from the regularity theory from Chap. 6. It states that only few of the projections contribute significantly to a solution of the electronic Schrödinger equation and estimates the speed of convergence of the expansion (9.34) in terms of the angular momentum quantum numbers ℓ_i. In analogy to Sect. 7.3 let u_ε denote that part of the expansion that is built up from the contributions assigned to the multi-indices ℓ for which

$$\prod_{i \in I_-} \left(1 + \ell_i(\ell_i + 1)\right) + \prod_{i \in I_+} \left(1 + \ell_i(\ell_i + 1)\right) < \frac{1}{\varepsilon^2}. \tag{9.40}$$

Since the decomposition (9.34) is orthogonal with respect to the H^1-norm then

$$\|u - u_\varepsilon\|_1 \le \varepsilon \, \|u - u_\varepsilon\| \le \varepsilon \, \|u\|. \tag{9.41}$$

9.3 The Radial Schrödinger Equation

Theorem 9.9 expresses a new kind of regularity that is not reflected in the results for the eigenfunctions expansions from Chap. 7. It can serve to further reduce the set of the eigenfunctions to be taken into account in such expansions and in most cases to improve the convergence rate. Before we can study this, we have to return to the eigenvalue problem for the three-dimensional Schrödinger-like operators

$$H = -\Delta + V \tag{9.42}$$

considered there with locally square integrable, nonnegative potentials V tending to infinity, but now under the restriction that these potentials are rotationally symmetric and infinitely differentiable outside the origin. The solution space $\mathcal{H}$ of such an eigenvalue problems is the completion of the space of the infinitely differentiable functions with compact support under the norm induced by the inner product

$$a(u,v) = \int \left\{\nabla u \cdot \nabla v + V u v\right\} dx, \tag{9.43}$$

a norm that dominates the L_2-norm and the H^1-norm as follows from the positivity of the eigenvalues. The eigenvalue problem for such operators splits into one-dimensional eigenvalue problems for the radial parts of the eigenfunctions, the reason being the following observation, which is a simple consequence of the rotational symmetry of the problem and the properties of the spherical harmonics:

Lemma 9.10. *For all infinitely differentiable functions u and v with compact support*

$$a(u,v) = \sum_{\ell=0}^{\infty} \sum_{m=-\ell}^{\ell} a(Q_\ell^m u, Q_\ell^m v). \tag{9.44}$$

The projections Q_ℓ^m are moreover symmetric in the sense that for these u and v

$$a(Q_\ell^m u, v) = a(u, Q_\ell^m v). \tag{9.45}$$

Proof. The relation (9.44) can be proved considering the two parts of which the inner product is composed separately. The first part involving the derivatives has been treated in Theorem 9.3. As V is rotationally symmetric, $Q_\ell^m V u = V Q_\ell^m u$. The rest follows by decomposing the square integrable functions Vu and v,

$$(Vu, v) = \sum_{\ell=0}^{\infty} \sum_{m=-\ell}^{\ell} (Q_\ell^m V u, Q_\ell^m v) = \sum_{\ell=0}^{\infty} \sum_{m=-\ell}^{\ell} (V Q_\ell^m u, Q_\ell^m v),$$

which is permissible because of the local square-integrability of the potential. The symmetry is proved by the same type of arguments.   □

The first equation shows that the operators Q_ℓ^m can be extended to projectors that are defined on the whole solution space $\mathcal{H}$ of the eigenvalue problem and are not only orthogonal with respect of the L_2-norm and the H^1-norm but also with respect to the norm that is induced by the inner product (9.43). The relations (9.44) and (9.45) thus transfer to all functions u and v in $\mathcal{H}$, a fact that allows us to decompose the eigenfunctions into radial and angular parts:

Theorem 9.10. *Let $u \neq 0$ be an eigenfunction for the isolated eigenvalue λ of finite multiplicity, that is, a function in the Hilbert space $\mathcal{H}$ satisfying the relation*

$$a(u, v) = \lambda(u, v), \quad v \in \mathcal{H}. \tag{9.46}$$

Then only finitely many of its projections $Q_\ell^m u$ are different from zero, and each of them is an eigenfunction for the eigenvalue λ too, that is, for all functions $v \in \mathcal{H}$

$$a(Q_\ell^m u, v) = \lambda(Q_\ell^m u, v). \tag{9.47}$$

The proof is a simple consequence of the fact that $(u, Q_\ell^m v) = (Q_\ell^m u, v)$ for all functions u and v in L_2, the lemma, and the finite dimension of the eigenspace. The remaining projections $Q_\ell^m u \neq 0$ hence span the eigenspace for the given eigenvalue. The original problem thus splits into the essentially one-dimensional eigenvalue problems to find functions u in the ranges of the projectors Q_ℓ^m satisfying the relation (9.46) for all test functions v in these subspaces. The resulting equation is the weak form of the radial Schrödinger equation.

Lemma 9.11. *The range of the projectors Q_ℓ^m on $\mathcal{H}$ consists of the functions*

$$u(x) = \frac{1}{r} f(r) K_\ell^m(x), \quad r = |x|, \tag{9.48}$$

whose radial parts are located in the completion of the space of the infinitely differentiable functions $f : \mathbb{R}_{\geq 0} \to \mathbb{R}$ with bounded support that vanish on a neighborhood of the point $r = 0$ under the norm given by the expression

$$\|f\|^2 = \int_0^\infty \left(f'(r)^2 + \frac{\ell(\ell+1)}{r^2} f(r)^2 + V(r) f(r)^2 \right) dr. \tag{9.49}$$

Proof. We observe first that the infinitely differentiable functions $u : \mathbb{R}^3 \rightarrow \mathbb{R}$ with compact support that vanish on a neighborhood of the origin form a dense subset of the solution space $\mathscr{H}$. That is again proved multiplying the infinitely differentiable functions with compact support with a series of cut-off functions χ_ε as in the proof of Theorem 9.2, using that the function $r \rightarrow 1/r^2$ is locally integrable in three space dimensions. The image of these functions under Q_ℓ^m are the functions (9.48) with f an infinitely differentiable function with bounded support vanishing on a neighborhood of the point $r = 0$. For such functions u and v

$$a(u,v) = \int \{-\Delta u + Vu\} v \, dx.$$

Let f and g be the radial parts of u and v. Since $x \cdot \nabla H_\ell^m = \ell H_\ell^m$ and $\Delta H_\ell^m = 0$ holds for the polynomials H_ℓ^m associated to the spherical harmonics K_ℓ^m,

$$(\Delta u)(x) = \frac{1}{r} \left(f''(r) - \frac{\ell(\ell+1)}{r^2} f(r) \right) K_\ell^m(x).$$

Inserting this relation above and integrating by parts one obtains the representation

$$a(u,v) = \int_0^\infty \left(f'(r)g'(r) + \frac{\ell(\ell+1)}{r^2} f(r)g(r) + V(r)f(r)g(r) \right) dr$$

of the inner product of such functions u and v, which proves the proposition. $\square$

We can conclude from the representation of the bilinear form (9.43) on the range of the operators Q_ℓ^m which we found in the proof above that the radial parts of the eigenfunctions in this range are weak solutions of the ordinary differential equation

$$-f''(r) + \frac{\ell(\ell+1)}{r^2} f(r) + V(r)f(r) = \lambda f(r), \quad r > 0. \qquad (9.50)$$

These differential equations do not depend on m, which means that the eigenvalues have at least the multiplicity $2\ell + 1$ and are degenerate except for the case $\ell = 0$.

Lemma 9.12. *The functions f that can be approximated arbitrarily well in the H^1-sense by infinitely differentiable functions f_n with compact support in the interval $r > 0$ are continuous and vanish at $r = 0$.*

Proof. The proposition follows from the estimate

$$|f_n(r) - f_m(r)|^2 \leq R \int_0^R |f_n'(s) - f_m'(s)|^2 \, ds,$$

that is proven with help of the fundamental theorem of calculus and the Cauchy-Schwarz inequality. Convergence in the H^1-norm implies thus uniform convergence on every interval $0 \leq r \leq R$ and with that continuity of the limit function. $\square$

To characterize the ranges of the projection operators Q_ℓ^m and the solutions of the radial Schrödinger equation further, we need the following Hardy-type inequality:

Lemma 9.13. *Let $f : \mathbb{R}_{\geq 0} \to \mathbb{R}$ be a continuous, square integrable function that is continuously differentiable on the interval $r > 0$, let its derivative be square integrable, and let $f(0) = 0$. The function $r \to f(r)/r$ is then square integrable and*

$$\int_0^\infty \frac{1}{r^2} f(r)^2 \, dr \leq 4 \int_0^\infty f'(r)^2 \, dr. \tag{9.51}$$

Proof. We can assume that f vanishes outside some bounded interval; otherwise one multiplies f again with a series of cut-off functions. Let $\delta > 0$ be arbitrary. Then

$$\int_\delta^\infty \frac{1}{r^2} f(r)^2 \, dr = \frac{1}{\delta} f(\delta)^2 + 2 \int_\delta^\infty \frac{1}{r} f(r) f'(r) \, dr,$$

as is shown integrating by parts. Because of $2ab \leq a^2/2 + 2b^2$,

$$2 \int_\delta^\infty \frac{1}{r} f(r) f'(r) \, dr \leq \frac{1}{2} \int_\delta^\infty \frac{1}{r^2} f(r)^2 \, dr + 2 \int_\delta^\infty f'(r)^2 \, dr.$$

Inserting this above one obtains from that the estimate

$$\int_\delta^\infty \frac{1}{r^2} f(r)^2 \, dr \leq \frac{2}{\delta} f(\delta)^2 + 4 \int_\delta^\infty f'(r)^2 \, dr.$$

To estimate the first term on the right hand side, let $0 < \varepsilon < \delta$. The fundamental theorem of calculus and the Cauchy-Schwarz inequality yield

$$|f(\delta)| = \left| f(\varepsilon) + \int_\varepsilon^\delta f'(r) \, dr \right| \leq |f(\varepsilon)| + \sqrt{\delta} \left(\int_0^\delta f'(r)^2 \, dr \right)^{1/2}$$

or, because of $f(0) = 0$ and the continuity of f at $r = 0$, in the limit as $\varepsilon \to 0+$

$$\frac{1}{\delta} f(\delta)^2 \leq \int_0^\delta f'(r)^2 \, dr.$$

Inserting this estimate above and letting δ tend to zero the proposition follows. □

Lemma 9.13 shows that the norms given by the expression (9.49) are all equivalent and that that for the index $\ell = 0$ does not play a special role. Moreover:

Lemma 9.14. *A function $f : \mathbb{R}_{\geq 0} \to \mathbb{R}$ that is continuous, infinitely differentiable on the interval $r > 0$, that vanishes at $r = 0$, and for which the expression (9.49) remains finite, can be approximated arbitrarily well by infinitely differentiable functions with compact support in the interval $r > 0$ in the sense of the norm given by (9.49).*

Proof. The idea is to multiply f by sequences of cut-off functions. We begin with the origin. Let $\chi : \mathbb{R} \to [0,1]$ be an infinitely differentiable function that takes the values $\chi(r) = 0$ for $r \leq 1$ and $\chi(r) = 1$ for $r \geq 2$. Set $\chi_\varepsilon(r) = \chi(r/\varepsilon)$ for $\varepsilon > 0$. As

$$|(\chi_\varepsilon f)(r)|^2 \leq |f(r)|^2, \quad |(\chi_\varepsilon f)'(r)|^2 \lesssim |f'(r)|^2 + \frac{1}{r^2} |f(r)|^2,$$

with some constant independent of ε in the second inequality, and

$$\lim_{\varepsilon \to 0+} (\chi_\varepsilon f)(r) = f(r), \quad \lim_{\varepsilon \to 0+} (\chi_\varepsilon f)'(r) = f'(r)$$

for all $r > 0$, the functions $\chi_\varepsilon f$ tend to f in the sense of the norm given by (9.49), as one shows in the usual way with help of the dominated convergence theorem. The square-integrability of the function $f(r)/r$ on the right hand side of the estimate above needed for that follows from the previous lemma. Let f now already vanish in a neighborhood of the point $r = 0$. The functions

$$r \to \left(1 - \chi\left(\frac{r}{R}\right)\right) f(r)$$

vanish then for $r \geq 2R$ and tend to f as R goes to infinity. $\square$

After these preparations we can now prove the second central result of this section:

Theorem 9.11. *The solutions of the radial Schrödinger equation (9.47) are functions of the form (9.48) with radial parts $f : \mathbb{R}_{\geq 0} \to \mathbb{R}$ that are continuous, that are infinitely differentiable on the interval $r > 0$ and solve there the differential equation (9.50) in the classical sense, that vanish at $r = 0$, and for which the expression (9.49) remains finite. They are completely characterized by these properties.*

Proof. Let $f : \mathbb{R}_{\geq 0} \to \mathbb{R}$ be the radial part of a weak solution of the radial Schrödinger equation (9.47). By Lemma 9.11 and Lemma 9.12, f is continuous on its interval of definition and vanishes at $r = 0$. We assign the function

$$\phi(r) = \lambda f(r) - \frac{\ell(\ell+1)}{r^2} f(r) - V(r) f(r)$$

to f that is then, under the given conditions on V, continuous on the interval $r > 0$. The function f itself solves the equation $-f'' = \phi$ in the weak sense, that is,

$$\int f'(r) \chi'(r)\, dr = \int \phi(r) \chi(r)\, dr$$

holds for all infinitely differentiable χ with compact support in the interval $r > 0$. Another weak solution of this equation is the solution g of the differential equation

$$-g''(r) = \phi(r), \quad r > 0,$$

that is fixed, say, by the values $g(1) = 0$ and $g'(1) = 0$. The difference $h = f - g$ is then a weak solution of the equation $-h'' = 0$ on the interval $r > 0$ and therefore a linear function. This can be shown with help of smoothed variants

$$(\delta_k * h)(r) = \int \delta_k(r - s) h(s)\, ds, \quad r > 1/k,$$

of h as they were considered in Sect. 2.3. Their second order derivatives

$$(\delta_k * h)''(r) = \int \delta_k'(r - s) h'(s)\, ds$$

vanish. They are therefore linear functions. Since they converge on every compact subinterval of the interval $r > 0$ to h in the L_1-sense as k goes to infinity, the limit function h is itself linear and $f = g + h$ therefore a twice continuously differentiable solution of the differential equation (9.50). Since V is not only continuous, as needed until now, but even infinitely differentiable, the solution is infinitely differentiable.

Conversely, that a solution f with the given properties can be approximated arbitrarily well by infinitely differentiable functions with compact support in the interval $r > 0$ in the sense of the norm given by (9.49) follows from Lemma 9.14.   □

The argument simplifies a little bit if $\ell \neq 0$, as the square integrability of the functions $f(r)/r$ follows then directly from the presence of the centrifugal barrier and does not need to be shown by arguments as in the proof of Lemma 9.13.

9.4 An Excursus to the Coulomb Problem

Before we continue with the study of the multi-particle case and of the approximation of high-dimensional wave functions we use the opportunity to calculate the hydrogen-like wave functions, the weak solutions of the Schrödinger equation

$$-\frac{1}{2}\Delta u - \frac{Z}{|x|} u = \lambda u \tag{9.52}$$

for a single electron in the field of a nucleus of charge Z. The knowledge about these eigenfunctions is basic for the qualitative understanding of chemistry and explains the structure of the periodic table to a large extent. These eigenfunctions have first been calculated by Schrödinger (with some help of Hermann Weyl) in his seminal article [73] that marks together with the work of Heisenberg the begin of modern quantum theory. The framework that has been developed in the previous section can be easily adapted to the given problem. We know from Chap. 4 that its solution space is the Sobolev space H^1. The eigenfunctions for a given eigenvalue λ are the linear combinations of functions of the form

$$u(x) = \frac{1}{r} f(r) K_\ell^m(x), \quad r = |x|, \tag{9.53}$$

with infinitely differentiable radial parts $f : \mathbb{R}_{>0} \to \mathbb{R}$ for which the expressions

$$\int_0^\infty f(r)^2 \, dr, \quad \int_0^\infty \left(f'(r)^2 + \frac{\ell(\ell+1)}{r^2} f(r)^2 \right) dr \tag{9.54}$$

representing the L_2-norm and the H^1-seminorm of u remain finite, which can be continuously extended by the value $f(0) = 0$ to $r = 0$, and which solve the equation

$$\frac{1}{2}\left(-f'' + \frac{\ell(\ell+1)}{r^2} f \right) - \frac{Z}{r} f = \lambda f. \tag{9.55}$$

Theorem 6.10, the virial theorem, shows that there are no nonnegative eigenvalues. We first show that there exist, for every given value $\lambda < 0$, exactly one solution of the equation (9.55) that tends to zero as r goes to zero, of course up to a multiplicative constant, and that this solution is even an entire function. The idea behind this ansatz is that the solutions of the equation (9.55) should for large r essentially behave like the square integrable solutions of the simplified equation in which the Coulomb term and the centrifugal term with $1/r$ and $1/r^2$ in front are neglected.

Lemma 9.15. *The only solutions of the differential equation (9.55) that can be continuously extended by the value $f(0) = 0$ to $r = 0$ are the multiples of the function*

$$f(r) = \phi(2\gamma r)\exp(-\gamma r), \quad \gamma = \sqrt{-2\lambda}, \tag{9.56}$$

where the leading factor is, up to the rescaling of the variable, the entire function

$$\phi(z) = z^{\ell+1} \sum_{k=0}^\infty a_k z^k \tag{9.57}$$

whose coefficients are normalized by the condition $a_0 = 1$ and satisfy the recursion

$$a_{k+1} = \frac{(k+\ell+1) - v}{(k+\ell+1)(k+\ell+2) - \ell(\ell+1)} a_k, \quad v = \frac{Z}{\gamma}. \tag{9.58}$$

Proof. The function (9.56) is a solution of the differential equation (9.55) if and only if the function (9.57) solves the differential equation

$$\phi'' - \phi' - \frac{\ell(\ell+1)}{z^2} \phi + \frac{v}{z} \phi = 0,$$

which is achieved by the choice of the coefficients. As the series defining this function converges for all complex numbers z, we have found a solution of the original equation. It remains to show that, up to the multiplication with a constant, there is no other solution that can be continuously extended by the value 0 to $r = 0$. The given function (9.56) can be extended to a power series solution

$$f(z) = z^{\ell+1} \sum_{k=0}^{\infty} b_k z^k, \quad b_0 \neq 0,$$

of the complex counterpart of the differential equation (9.55). That means that there exists a $\delta > 0$ such that $z \to 1/f(z)^2$ possesses, for $0 < |z| < \delta$, a Laurent expansion

$$\frac{1}{f(z)^2} = \frac{1}{z^{2\ell+2}} \sum_{k=0}^{\infty} c_k z^k,$$

with a coefficient $c_0 \neq 0$. Let $r_0 = \delta/2$. The real-valued function

$$g(r) = c(r)f(r), \quad c(r) = \int_{r_0}^{r} \frac{1}{f(s)^2}\, ds,$$

is then well-defined for $0 < r < \delta$ and solves the differential equation (9.55) on this interval. As such it can be uniquely extended to a solution of this equation on the whole interval $r > 0$. Term-wise integration, permissible because of the uniform convergence of the series on compact subintervals of the interval $0 < r < \delta$, yields

$$c(r) = \sum_{k} \frac{c_k}{k - 2\ell - 1} r^{k-2\ell-1} + c_{2\ell+1} \ln r + \alpha,$$

where the sum extends over all nonnegative integers k except for $k = 2\ell + 1$ and α is an integration constant. The solution g thus behaves near the point $r = 0$ up to a non-vanishing, otherwise uninteresting constant factor like $\sim 1/r^{\ell}$. For $\ell = 0$ it tends to a value $\neq 0$ and for $\ell \geq 1$ it even becomes singular as r goes to zero. It is therefore linearly independent of f and spans together with f the solution space of the differential equation (9.55), which completes the proof. $\qquad \square$

The solutions (9.56) are the only candidates for the radial parts of the eigenfunctions. They lead to solutions (9.53) of the Schrödinger equation (9.52) if and only if the integrals (9.54) remain finite, or equivalently f and its first order derivative are square integrable. This is only the case for particular values of λ.

Lemma 9.16. *If the series (9.57) does not terminate the regular solution (9.56) of the differential equation (9.55) tends exponentially to infinity as r goes to infinity.*

Proof. If the series does not terminate there exists for each $\varepsilon < 1$ an n such that

$$\frac{a_{k+1}}{a_k} \geq \frac{1 - \varepsilon}{k}$$

holds for all indices for all $k \geq n$. From that one can derive a lower bound

$$|\phi(2\gamma r)| \gtrsim r^{\ell+1} \exp((1 - \varepsilon)2\gamma r) + p(r),$$

with p a polynomial, for the absolute value of the function ϕ. The absolute value of the solution (9.56) tends in this case exponentially to infinity.   □

Conversely, if the series terminates, the function (9.57) reduces to a polynomial. The exponential term on the right hand side of (9.56) then dominates and the function (9.56) and its derivative tend exponentially to zero. With that we have completely solved the Schrödinger equation (9.52). The eigenfunctions are linear combinations of the eigenfunctions (9.53) with radial parts (9.56). The eigenvalues are

$$\lambda = -\frac{Z^2}{2(n_r+\ell+1)^2}, \quad n_r = 0,1,2,\ldots. \tag{9.59}$$

They depend only on the principal quantum number $n = n_r+\ell+1$. For given principal quantum number $n = 1,2,\ldots$, the possible angular momentum quantum numbers are $\ell = 0,\ldots,n-1$, and for given n and ℓ, the possible magnetic quantum numbers $m = -\ell,\ldots,\ell$. The dimension of the eigenspace for the eigenvalue

$$\lambda = -\frac{Z^2}{2n^2} \tag{9.60}$$

is therefore n^2, so that the higher eigenvalues are highly degenerate. The eigenvalues cluster at the ionization threshold, the minimum of the essential spectrum. The eigenfunctions, in particular that for the minimum eigenvalue, exhibit singularities at the position of the nucleus typical for electronic wave functions.

9.5   The Harmonic Oscillator

For us the most important example of a three-dimensional Schrödinger operator (9.42) to which the considerations of Sect. 9.3 directly apply is the three-dimensional harmonic oscillator. Its eigenfunctions have already been determined in Sect. 3.4. Their completeness was the key to Theorem 9.1 so that a study based on the results of Sect. 9.3 cannot directly replace our former considerations but can give a much more detailed information about the structure of the eigenfunctions. We know from Sect. 9.3 that the solutions of the Schrödinger equation

$$-\frac{1}{2}\Delta u + \frac{1}{2}|x|^2 u = \lambda u \tag{9.61}$$

can be composed of solutions of the form

$$u(x) = \frac{1}{r} f(r) K_\ell^m(x), \tag{9.62}$$

with radial parts that solve the differential equation

$$\frac{1}{2}\left(-f'' + \frac{\ell(\ell+1)}{r^2}f\right) + \frac{1}{2}r^2 f = \lambda f. \tag{9.63}$$

This equation possesses again a power series solution that vanishes at the origin:

Lemma 9.17. *The only solutions of the differential equation (9.63) that can be continuously extended by the value $f(0) = 0$ to $r = 0$ are the multiples of the function*

$$f(r) = \phi(r)e^{-r^2/2}, \tag{9.64}$$

where the leading factor, in front of the exponential term, is the entire function

$$\phi(z) = z^{\ell+1}\sum_{k=0}^{\infty} a_k z^{2k} \tag{9.65}$$

whose coefficients are normalized by the condition $a_0 = 1$ and satisfy the recursion

$$a_{k+1} = \frac{(4k+2\ell+3) - 2\lambda}{(2k+\ell+2)(2k+\ell+3) - \ell(\ell+1)}\, a_k. \tag{9.66}$$

Proof. The function (9.64) is a solution of the differential equation (9.63) if and only if the function (9.65) solves the differential equation

$$\phi'' - 2z\phi' - \frac{\ell(\ell+1)}{z^2}\phi + (2\lambda - 1)\phi = 0.$$

The coefficients a_k are chosen accordingly. To exclude further solutions of the equation (9.63), up to constant multiples of the function (9.64), one can literally transfer the arguments from the proof of Lemma 9.15. Since $z \to 1/f(z)^2$ is an even function the logarithmic term there does not appear in the present case and the second solution can be expanded into a Laurent series in a neighborhood of the origin. □

The power series in (9.65) collapses to an even polynomial p_{2n} of order $2n$ if

$$\lambda = 2n + \ell + \frac{3}{2}, \quad n = 0,1,2,\ldots. \tag{9.67}$$

The functions (9.62) become then the polynomial multiples

$$u(x) = r^\ell p_{2n}(r) K_\ell^m(x) e^{-r^2/2} \tag{9.68}$$

of the Gauss function $r \to e^{-r^2/2}$ and are admissible solutions of the Schrödinger equation (9.61) for the eigenvalues (9.67). Since we know from Lemma 9.7 respectively from (9.18) that these solutions span all polynomial multiples of that Gaussian, and from Sect. 3.4 that all eigenfunctions are of this type, we can stop our considerations here and have separated the radial from the angular dependence.

9.6 Eigenfunction Expansions Revisited

In Chaps. 7 and 8 expansions of N-particle wave functions into tensor products
of one-particle eigenfunctions have been studied and estimates for their convergence rates have been given. These estimates are based on the regularity theory from
Chap. 6, that is, on the existence and the decay properties of the mixed derivatives of
the solutions of the electronic Schrödinger equation. However, they do not fully exploit these regularity properties of the solutions and will be refined in the present
section. As in Chap. 7 we start from the eigenfunctions of a three-dimensional
Schrödinger operator (7.1), but assume now as in Sect. 9.3 that the underlying potential $V \geq 0$ is not only locally square integrable and tends to infinity, but is also
rotationally symmetric and infinitely differentiable outside the origin. Let

$$\phi_{n\ell m}(x) = \frac{1}{r} f_{n\ell}(r) K_{\ell}^{m}(x), \quad n, \ell = 0, 1, 2, \ldots, \quad m = -\ell, \ldots, \ell, \tag{9.69}$$

be solutions of the three-dimensional Schrödinger equation

$$-\Delta \phi_{n\ell m} + V \phi_{n\ell m} = \lambda_{n\ell} \phi_{n\ell m} \tag{9.70}$$

as studied in Sect. 9.3, which are pairwise orthogonal, have L_2-norm 1, and span a
dense subspace of L_2. We consider in this section the orthogonal expansions

$$u(x) = \sum_{n, \ell, m} \widehat{u}(n, \ell, m) \prod_{i=1}^{N} \phi_{n_i \ell_i m_i}(x_i) \tag{9.71}$$

of square integrable functions u defined on the $(\mathbb{R}^3)^N$ into tensor products of these
eigenfunctions. The sum here runs over the complete set of these products, i.e., over
the multi-indices n, ℓ, and m with integer components $n_i, \ell_i \geq 0$ and $|m_i| \leq \ell_i$.

Up to here nothing has changed from Chap. 7, except for the labeling of the
eigenvalues and eigenfunctions of the three-dimensional operator. The point is that
the eigenfunctions (9.69) are also eigenfunctions of the angular momentum operator L^2 and that the projections $Q(\ell, m)u$ considered in Sect. 9.2 can in the present
case be easily expressed in terms of the given expansion. It is

$$(Q(\ell, m)u)(x) = \sum_{n} \widehat{u}(n, \ell, m) \prod_{i=1}^{N} \phi_{n_i \ell_i m_i}(x_i). \tag{9.72}$$

The angular parts are kept fixed and the sum extends only over the corresponding radial parts. At this place the results from Sect. 9.2 come into play, in particular
Theorem 9.9. Together with Theorem 7.1 and the considerations in Sect. 7.3 they imply that one can restrict oneself to contributions assigned to multi-indicesfor which

$$\prod_{i\in I_-} \left(1+\ell_i(\ell_i+1)\right) + \prod_{i\in I_+} \left(1+\ell_i(\ell_i+1)\right) < \frac{1}{\varepsilon^2}, \tag{9.73}$$

$$\prod_{i\in I_-} \frac{\lambda_{n_i\ell_i}}{\Omega^2} + \prod_{i\in I_+} \frac{\lambda_{n_i\ell_i}}{\Omega^2} < \frac{1}{\varepsilon^2} \tag{9.74}$$

to reach an H^1-approximation error of order $\mathcal{O}(\varepsilon)$ for the solutions of the electronic Schrödinger equation, provided the potential V is adapted to the considered eigenfunction as described in Sect. 7.1. The products run as always over the sets of the indices of the electrons with spin $-1/2$ and spin $+1/2$. The extension to the antisymmetric case considered in Sect. 7.2, in which the product of the eigenfunctions is replaced by a Slater determinant, respectively by the product of two Slater determinants, is obvious. The additional condition (9.73) can reduce the number of contributions to be taken into account substantially. Many of the regularity properties of the solutions that have not been utilized in Theorem 7.1 enter at this place.

One can often even go a step further and make use of the fact that the functions (9.69) are not only eigenfunctions of the operator (9.42) but also of the operators

$$H + \omega L^2 = -\Delta + V + \omega L^2, \tag{9.75}$$

whose eigenvalues are shifted from $\lambda_{n\ell}$ to $\lambda_{n\ell} + \ell(\ell+1)\omega$. We will demonstrate this in the next section by means of the eigenfunctions of the harmonic oscillator.

9.7 Approximation by Gauss Functions

The expansion of the solutions of the electronic Schrödinger equation into tensor products of eigenfunctions of the harmonic oscillator, that is, into the easily manageable Gauss functions, has already been discussed in Chap. 8. We found that the H^1-error decreases in the one-particle case at least like $\sim n^{-1/6}$ in the number n of the involved basis functions. Almost the same holds for the N-particle case if all particles have the same spin. We will improve this estimate for the convergence rate in this section to $\sim n^{-1/4}$, which comes much closer to the upper bound $\sim n^{-1/3}$. Let

$$H' = -\frac{1}{2}\Delta + \frac{1}{2}|x|^2 + L^2 = H + L^2 \tag{9.76}$$

be the Hamilton operator of the three-dimensional harmonic oscillator to which the square (9.2) of the angular momentum operator is added. Its eigenfunctions

$$\phi_{n\ell m}(x) = \frac{1}{r} f_{n\ell}(r) K_\ell^m(x), \quad n,\ell = 0,1,2,\ldots, \ m = -\ell,\ldots,\ell, \tag{9.77}$$

are those of the harmonic oscillator. The eigenvalues assigned to them are

$$\lambda'_{n\ell} = 2n+\ell+\frac{3}{2}+\ell(\ell+1) = \lambda_{n\ell}+\ell(\ell+1). \tag{9.78}$$

Next we introduce a new norm on the space $\mathscr{D}$ of the infinitely differentiable functions from $\mathbb{R}^{3N}$ to $\mathbb{R}$ with compact support. It is given by the expression

$$\||u|\|^2 = \left(\left(\sum_{i=1}^N H_i \right) \left(\prod_{i\in I_-} H_i' + \prod_{i\in I_+} H_i' \right) u, u \right), \qquad (9.79)$$

where the H_i, respectively H_i', are as in Sect. 7.1 the counterparts of the three-dimensional operators (9.76) that act on the coordinates of the electron i. Distributing the derivatives in equal parts to both sides of the inner product, one can estimate this norm by the L_2-norms of polynomially weighted derivatives of u as they have been considered in Theorem 6.13. The solutions of the electronic Schrödinger equation for eigenvalues below the essential spectrum are therefore contained in the completion of $\mathscr{D}$ under the norm given by (9.79). On the other hand,

$$\||u|\|^2 = \sum_{n,\ell,m} \left(\sum_{i=1}^N \lambda_{n_i\ell_i} \right) \left(\prod_{i\in I_-} \lambda_{n_i\ell_i}' + \prod_{i\in I_+} \lambda_{n_i\ell_i}' \right) |\widehat{u}(n,\ell,m)|^2, \qquad (9.80)$$

first for infinitely differentiable functions u with compact support. That means that a square integrable function u belongs to the completion of $\mathscr{D}$ under the norm given by (9.79) if and only if the expression (9.80) remains finite. In particular this holds for the solutions of the electronic Schrödinger equation.

To approximate the solutions of the electronic Schrödinger equation up to an H^1-error of order $\mathscr{O}(\varepsilon)$ it suffices therefore again to restrict to the contribution of the tensor products of the eigenfunctions (9.77) for which the estimate

$$\prod_{i\in I_-} \lambda_{n_i\ell_i}' + \prod_{i\in I_+} \lambda_{n_i\ell_i}' < \frac{1}{\varepsilon^2} \qquad (9.81)$$

holds for the assigned eigenvalues. We need therefore to know how fast the eigenvalues (9.78) increase to determine the convergence rate in terms of the number of the remaining, correspondingly antisymmetrized tensor products.

Lemma 9.18. *The number $\mathscr{N}(\lambda)$ of the eigenvalues $\lambda_{n\ell}' < \lambda$, counted with their multiplicity, behaves asymptotically like $\sim \lambda^2/4$ and is bounded by the expression*

$$\mathscr{N}(\lambda) \le \frac{5}{2}\lambda^2. \qquad (9.82)$$

Proof. Let $L = \max\{\ell \mid (\ell+1)^2 + 1/2 < \lambda\}$. Since $\lambda_{n\ell}' < \lambda$ if and only if $\ell \le L$ and

$$n < \frac{\lambda}{2} - \frac{(\ell+1)^2}{2} - \frac{1}{4} \le \frac{(L+2)^2}{2} - \frac{(\ell+1)^2}{2},$$

the number $\mathscr{N}(\lambda)$ of the eigenvalues $\lambda_{n\ell}' < \lambda$ is bounded by the sum

$$\mathcal{N}(\lambda) \leq \sum_{\ell=0}^{L} \left(\frac{(L+2)^2}{2} - \frac{(\ell+1)^2}{2} + 1 \right)(2\ell+1).$$

This sum behaves asymptotically like $\sim L^4/4$, i.e., like $\sim \lambda^2/4$, and is bounded by

$$\mathcal{N}(\lambda) \leq \frac{5}{2}(L+1)^4.$$

Since $(L+1)^2 \leq \lambda$ this proves the estimate (9.82). Conversely the lower estimate

$$\mathcal{N}(\lambda) \geq \sum_{\ell=0}^{L} \left(\frac{(L+1)^2}{2} - \frac{(\ell+1)^2}{2} \right)(2\ell+1),$$

holds. This bound behaves asymptotically like $\sim L^4/4$ i.e., like $\sim \lambda^2/4$ as well. $\quad\square$

We label the eigenvalues $\lambda'_{n\ell}$ now as in the previous chapters by a single index k and order them ascendingly, where they are counted several times according to their multiplicity. We can then conclude from (9.82) that they increase at least like

$$\lambda'_k \geq \sqrt{\frac{2}{5}k}. \tag{9.83}$$

That is considerably more rapid than the growth $\lambda_k \sim k^{1/3}$ of the eigenvalues of the harmonic oscillator itself that has been studied in Lemma 8.1.

From here we can proceed as in Chap. 8. For the case that all electrons have the same spin, the number of antisymmetrized tensor products or Slater determinants built from the eigenfunctions (9.77) that are needed to reach an H^1-error of order $\mathcal{O}(\varepsilon)$ increases essentially like $\mathcal{O}(\varepsilon^{-4})$ for ε tending to zero, independent of the number of the electrons. The convergence rate improves by that by two orders compared to the estimate from Chap. 8. This reflects the fact that the Gauss functions behave in angular direction like the eigenfunctions of any other operator (9.42), including those with much more rapidly increasing potentials and therefore more rapidly increasing eigenvalues. The reduction of the convergence order observed in Chap. 8 is exclusively due to the radial behavior of the Gauss functions. In the general case of electrons of distinct spin, the order of convergence of the hyperbolic cross approximation halves, again due to the singularities of the wave functions at the places where electrons with opposite spin meet.

9.8 The Effect of Scaling

Gauss functions have a lot of attractive features far beyond the convergence properties just discussed that are remarkable but do not fully explain their success. The first reason for the popularity and the almost exclusive use of Gauss functions in quantum chemistry is the observation due to Boys [12] that the integrals arising in every variational procedure to solve the Schrödinger equation attain a comparatively simple form and can be evaluated more easily than with other basis sets. A second reason is their scaling and invariance properties. Gauss functions

$$x \;\rightarrow\; P(x)\,\mathrm{e}^{-|x|^2/2} \tag{9.84}$$

are products of a fixed exponential part with polynomials. Polynomials remain polynomials of same degree under any kind of linear transformation, under a rescaling of the variables in the same way as under rotations, shears, or shifts. Ansatz spaces like the given hyperbolic cross spaces become scaling invariant approximation manifolds if one allows for a scaling of the exponential part, either individually for each single electron or jointly for all. Such measures can improve the approximation properties dramatically and enhance the speed of convergence substantially.

As an example we consider the hydrogen orbitals that have been calculated in Sect. 9.4. The first observation is that the angular parts of the Gauss functions coincide with the angular parts of the hydrogen orbitals since the Hamiltonians of both problems are rotationally invariant. The angular parts can therefore be kept fixed and only the radial parts need to be approximated by a linear combination of the radial parts of the corresponding Gauss functions. Our estimates guarantee that the H^1-error tends to zero in this case at least like $\sim n^{-1/2}$, and the error of the eigenvalues at least like $\sim n^{-1}$, in the number n of the included Gaussians. Such convergence orders might be acceptable for a basically three-dimensional problem but are surely not overwhelming in view of the fact that the symmetry properties of the problem are here taken into account explicitly.

The situation changes immediately if one allows for a rescaling of the exponential parts of the Gaussians and combines several such rescaled Gauss functions. This can be recognized as follows. The construction starts from the representation

$$\mathrm{e}^{-\sqrt{s}} = \int_0^\infty F(t)\,\mathrm{e}^{-st}\,\mathrm{d}t, \quad F(t) = \frac{1}{2\sqrt{\pi t^3}}\,\exp\!\left(-\frac{1}{4t}\right), \tag{9.85}$$

of the function $s \rightarrow \mathrm{e}^{-\sqrt{s}}$ on the interval $s \geq 0$ as Laplace transform, that one can take from mathematical tables like [1] or easily calculate with help of computer algebra programs. From (9.85) one obtains the integral representations

$$\mathrm{e}^{-r} = \int_0^1 g(\xi, r^2)\,\mathrm{d}\xi, \quad g(\xi, s) = F(t(\xi))\exp(-st(\xi))\,t'(\xi), \tag{9.86}$$

of the exponential function e^{-r}, where the functions $t(\xi)$ map the interval $0 < \xi < 1$ onto the positive real axis $t > 0$. We consider in the following the substitutions

$$t(\xi) = \frac{\xi^2}{(1-\xi)^{2m}}, \tag{9.87}$$

where the exponent $m \geq 2$ can be used to influence the properties of the integrand.

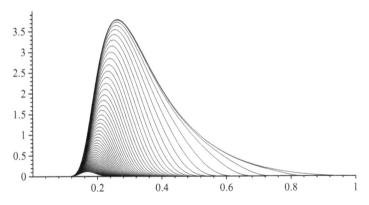

Fig. 9.1 The integrand in (9.86) for $m = 3$ and $r = 0.0$, $r = 0.01$, $r = 0.05$, $r = 0.1, 0.2, \ldots, 5.0$

Lemma 9.19. *The j-th derivative of the integrand $g(\xi, s)$ with respect to ξ remains uniformly bounded in $s \geq 0$ if $j \leq m - 1$, and uniformly in $s \geq s_0 > 0$ for arbitrary j. The integrand tends to zero at the boundary points $\xi = 0$ and $\xi = 1$. If $s > 0$ the same holds for all its derivatives at both boundary points.*

Proof. We start from the function $\alpha : [0, 1] \to \mathbb{R}$ that is for $0 < \xi \leq 1$ defined by

$$\alpha(\xi) = \frac{2 + (2m - 2)\xi}{2\sqrt{\pi}\,\xi^2} \exp\left(-\frac{(1 - \xi)^{2m}}{4\xi^2}\right)$$

and for $\xi = 0$ by $\alpha(0) = 0$. It is infinitely differentiable on the whole interval $[0, 1]$. All its derivatives take the value $\alpha^{(j)}(0) = 0$ at the left boundary point. One has

$$g(\xi, s) = (1 - \xi)^{m-1} \alpha(\xi) \exp(-st(\xi)).$$

The derivatives of $g(\xi, s)$ with respect to ξ can therefore be written in the form

$$g^{(j)}(\xi, s) = (1 - \xi)^{m-1-j} \sum_{k=0}^{j} \alpha_{jk}(\xi)(st(\xi))^k \exp(-st(\xi)).$$

The coefficient functions α_{jk} are infinitely differentiable on the interval $0 \leq \xi \leq 1$. Their derivatives vanish at $\xi = 0$ as those of α. The proof uses the representation

$$t'(\xi) = \frac{2 + (2m - 2)\xi}{\xi(1 - \xi)} t(\xi)$$

of the derivative of the function (9.87) and the fact that the functions $\alpha_{jk}(\xi)$ absorb every negative power $\xi^{-\nu}$ of ξ. Since the functions $x \to x^k e^{-x}$ are bounded on the interval $x \geq 0$ the representation above proves that the functions $\xi \to g^{(j)}(\xi, s)$ remain uniformly bounded in $s \geq 0$ as long as $j \leq m - 1$. Since

$$(1 - \xi)^{m-j-1} = s^{-\nu} \xi^{-2\nu} (st(\xi))^\nu, \quad \nu = \frac{j-m+1}{2m},$$

they remain uniformly bounded in $s \geq s_0 > 0$ for all $j \geq m$. If $s > 0$ the functions $\xi \to g(\xi, s)$ and all their derivatives tend to zero as ξ goes to 1. $\square$

These properties enable us to approximate the integrals (9.86) by the trapezoidal rule, that is, to approximate the exponential function $r \to e^{-r}$ by the functions

$$f_n(r) = \frac{1}{n+1} \sum_{k=1}^n g\left(\frac{k}{n+1}, r^2\right). \tag{9.88}$$

These functions converge uniformly to e^{-r}, at least like $\sim 1/n^{m-1}$ on the whole interval $r \geq 0$ and faster than any power of $1/n$ on all subintervals $r \geq r_0$ of the interval $r > 0$. This follows from the exactness of the trapezoidal rule for trigonometric polynomials of period 1 and degree n and the Fourier series representation of the L_2-norms of the derivatives of infinitely differentiable 1-periodic functions. The functions (9.88) are linear combinations of the Gauss functions

$$r \to \exp\left(-t\left(\frac{k}{n+1}\right) r^2\right), \quad k = 1, \ldots, n, \tag{9.89}$$

that depend only indirectly on the function to be approximated. If $m = 3$ is set the Rayleigh quotient takes in the functions $u_n(x) = f_n(r)$ approximating the ground state eigenfunction $u(x) = e^{-r}$ of the hydrogen atom for the given n the values

$$\begin{aligned}
n &= 16: \quad -0.49996499582807 \\
n &= 32: \quad -0.49999999906702 \\
n &= 64: \quad -0.49999999999999
\end{aligned}$$

that approach the exact ground state energy $-1/2$ very rapidly. These observations indicate that an astonishingly small number of Gauss functions already suffices to reach a high accuracy. In fact, Braess [13] and Kutzelnigg [55] have shown that one can reach even a kind of exponential convergence with linear combinations of such rescaled Gauss functions in the approximation of the hydrogen ground state. With good cause quantum chemistry today is largely based on the use of Gauss functions.

Appendix: The Standard Basis of the Spherical Harmonics

The aim of this appendix is to construct a basis of the space of the complex-valued spherical harmonics of degree ℓ and to continue the study of the radial-angular decomposition from Sect. 9.1. The first observation is that the complex vector space of the homogeneous harmonic polynomials of degree ℓ, and with that the space of these spherical harmonics, have the same dimension $2\ell + 1$ as their real counterparts. The proof is identical to that of Lemma 9.6. We start introducing the components

$$L_1 = -i\,(x_2 D_3 - x_3 D_2), \quad L_2 = -i\,(x_3 D_1 - x_1 D_3),$$
$$L_3 = -i\,(x_1 D_2 - x_2 D_1)$$

of the angular momentum operator $L = -i\,x \times \nabla$. Its square (9.2) can be expressed as

$$L^2 = L_1^2 + L_2^2 + L_3^2$$

in terms of these operators. Their commutators $[L_\nu, L_\mu] = L_\nu L_\mu - L_\nu L_\mu$ are

$$[L_1, L_2] = i\,L_3, \quad [L_2, L_3] = i\,L_1, \quad [L_3, L_1] = i\,L_2.$$

The operators L_1, L_2, and L_3 are formally self-adjoint with respect to the L_2-inner product on the space of the infinitely differentiable functions with compact support.

Lemma. *If H is a homogeneous harmonic polynomial, the functions $L_1 H$, $L_2 H$, and $L_3 H$ are homogeneous harmonic polynomials of the same degree.*

Proof. They are obviously homogeneous polynomials of the same degree. Since

$$\Delta\,(x_i D_j - x_j D_i) = (x_i D_j - x_j D_i)\,\Delta$$

they solve, like H, the Laplace equation and are harmonic.   □

Lemma. *Let H be a homogeneous harmonic polynomial of degree ℓ whose L_2-norm over the unit sphere is 1 and let $L_3 H = m H$. The harmonic polynomial*

$$H_- = L_- H, \quad L_- = L_1 - i\,L_2,$$

satisfies then the equation $L_3 H_- = (m-1)H_-$. Its L_2-norm over the unit sphere S is

$$\int_S |H_-(\eta)|^2\,d\eta \;=\; \ell(\ell+1) - m(m-1).$$

Proof. As follows from the commutation relations, $L_3 L_- = L_-(L_3 - 1)$. Thus

$$L_3 H_- = L_3 L_- H = (m-1)L_- H = (m-1)H_-.$$

To calculate the L_2-norm of H_- over the unit sphere S let $f \neq 0$ be a rotationally symmetric, infinitely differentiable function with compact support and set

$$\psi(x) = f(x)H(x), \quad \psi_-(x) = f(x)H_-(x).$$

As shown in the proof of Lemma 9.2 then $\psi_- = L_-\psi$ and

$$L^2\psi = fL^2H = \ell(\ell+1)fH = \ell(\ell+1)\psi,$$
$$L_3\psi = fL_3H = mfH = m\psi.$$

The formal self-adjointness of L_1 and L_2 and the commutation relations above yield

$$\|L_-\psi\|_0^2 = (\psi, (L^2 - L_3^2 + L_3)\psi) = (\ell(\ell+1) - m(m-1))\|\psi\|_0^2.$$

The L_2-norm of $H_- = L_-H$ over the unit sphere can be calculated from that using the homogeneity of H_- and H, in the same way as in the proof of Lemma 9.5.   □

Particularly, $H_- \neq 0$ as long m is different from $\ell+1$ and $-\ell$. Based on these observations it is now easy to construct the desired basis recursively, starting from the polynomial $z^\ell = (x_1 + i x_2)^\ell$ in the complex variable $z = x_1 + i x_2$.

Theorem. Let H_ℓ^ℓ be the homogeneous harmonic polynomial

$$H_\ell^\ell(x) = N_{\ell\ell}(x_1 + i x_2)^\ell, \quad N_{\ell\ell}^{-2} = 2\pi \int_{-\pi/2}^{\pi/2}(\cos\vartheta)^{2\ell+1}d\vartheta,$$

and let $N_{\ell m}^{-2} = \ell(\ell+1) - m(m-1)$. The polynomials H_ℓ^ℓ and

$$H_\ell^{m-1}(x) = N_{\ell m}(L_- H_\ell^m)(x), \quad m = \ell, \dots, -\ell+1,$$

together form an orthonormal basis of the space of the homogeneous harmonic polynomials of degree ℓ in the sense of the L_2-inner product on the unit sphere. They are eigenfunctions of both L^2 and L_3:

$$L^2 H_\ell^m = \ell(\ell+1)H_\ell^m, \quad L_3 H_\ell^m = mH_\ell^m.$$

Proof. That H_ℓ^ℓ is a normed homogeneous harmonic polynomial of degree ℓ and $L_3 H_\ell^\ell = \ell H_\ell^\ell$ is easily checked. That the H_ℓ^m are normed and $L_3 H_\ell^m = mH_\ell^m$ follows from the previous lemma. That they are orthogonal to each other is shown as in the proof of Lemma 9.4 and 9.5. As the space of the homogeneous harmonic polynomials of degree ℓ has the dimension $2\ell+1$, the H_ℓ^m thus span this space.   □

Due to their inherent symmetries as joint eigenfunctions of the operators L^2 and L_3 the polynomials H_ℓ^m can best be represented in polar coordinates:

$$H_\ell^m(r\cos\varphi\cos\vartheta, r\sin\varphi\cos\vartheta, r\sin\vartheta) = r^\ell Y_\ell^m(\varphi, \vartheta),$$

where the angles range in the intervals $0 \leq \varphi \leq 2\pi$ and $|\vartheta| \leq \pi/2$. The functions

$$Y_\ell^m(\varphi, \vartheta) = H_\ell^m(\cos\varphi\cos\vartheta, \sin\varphi\cos\vartheta, \sin\vartheta)$$

represent the restrictions of the H_ℓ^m to the surface of the unit sphere. They form the standard basis of the three-dimensional spherical harmonics.

It is not especially difficult to calculate the functions $Y_\ell^m(\varphi, \vartheta)$ explicitly. The first and most important observation is that they factor into products of univariate trigonometric polynomials in the variables φ and ϑ.

Theorem. *The three-dimensional spherical harmonics can be written as*

$$Y_\ell^m(\varphi, \vartheta) = e^{im\varphi} P_\ell^m(\vartheta), \quad P_\ell^m(\vartheta) = H_\ell^m(\cos\vartheta, 0, \sin\vartheta).$$

Proof. Differentiation with respect to the variable φ leads to

$$\frac{\partial}{\partial\varphi} Y_\ell^m(\varphi, \vartheta) = i\,(L_3 H_\ell^m)(\cos\varphi\cos\vartheta, \sin\varphi\cos\vartheta, \sin\vartheta),$$

that is, because of $L_3 H_\ell^m = m H_\ell^m$, to the differential equation

$$\frac{\partial}{\partial\varphi} Y_\ell^m(\varphi, \vartheta) = im\, Y_\ell^m(\varphi, \vartheta)$$

in the variable φ for ϑ kept fixed and therefore to the desired representation

$$Y_\ell^m(\varphi, \vartheta) = e^{im\varphi} Y_\ell^m(0, \vartheta)$$

of $Y_\ell^m(\varphi, \vartheta)$ as product of two univariate trigonometric polynomials.   □

The recursion for the polynomials H_ℓ^m can be translated into a recursion for the ϑ-parts of the spherical harmonics $Y_\ell^m(\varphi, \vartheta)$. Direct calculation shows

$$e^{-i\varphi}\left(i\tan\vartheta\,\frac{\partial}{\partial\varphi} + \frac{\partial}{\partial\vartheta}\right) Y_\ell^m(\varphi, \vartheta) = (L_- H_\ell^m)(\cos\varphi\cos\vartheta, \sin\varphi\cos\vartheta, \sin\vartheta).$$

Starting from $P_\ell^\ell(\vartheta) = N_{\ell\ell}(\cos\vartheta)^\ell$ one obtains P_ℓ^m, $m = \ell - 1, \ldots, -\ell$, therefore via

$$P_\ell^{m-1}(\vartheta) = N_{\ell m}\left(-m\tan\vartheta\, P_\ell^m(\vartheta) + \frac{d}{d\vartheta} P_\ell^m(\vartheta)\right).$$

This relation also shows that the functions P_ℓ^m are real-valued.

The function given by the expression $H(x) = H_\ell^m(x_1, -x_2, x_3)$ is like H_ℓ^m itself a normed homogeneous harmonic polynomial of degree ℓ. Since

$$(L_3 H)(x) = -(L_3 H_\ell^m)(x_1, -x_2, x_3) = -m H(x),$$

it is at the same time an eigenfunction of the operator L_3 for the eigenvalue $-m$ and therefore a complex multiple $H = \varepsilon H_\ell^{-m}$, $|\varepsilon| = 1$, of the polynomial H_ℓ^{-m}. This implies $P_\ell^m = \varepsilon P_\ell^{-m}$, or, since the P_ℓ^m are real-valued,

$$P_\ell^{-m}(\vartheta) = \pm P_\ell^m(\vartheta).$$

The φ-independent function $Y_\ell^0(\varphi, \vartheta) = P_\ell^0(\vartheta)$ and the trigonometric polynomials

$$\sqrt{2} \sin(m\varphi) P_\ell^m(\vartheta), \quad \sqrt{2} \cos(m\varphi) P_\ell^m(\vartheta), \quad m = 1, \dots, \ell,$$

thus form a real-valued, orthonormal basis of the spherical harmonics of degree ℓ.

Similarly as in the real case considered in Sect. 9.1, we assign to every infinitely differentiable function $u : \mathbb{R}^3 \to \mathbb{C}$ the functions given by

$$(Q_\ell^m u)(x) = \left\{ \int_S u(r\eta) \overline{K_\ell^m(\eta)} \, d\eta \right\} K_\ell^m(x),$$

where again $r = |x|$ and the functions K_ℓ^m are the spherical harmonics assigned to the harmonic polynomials H_ℓ^m, now in cartesian coordinates. They are given by

$$K_\ell^m(x) = H_\ell^m\left(\frac{x}{r}\right), \quad r = |x|.$$

The functions $Q_\ell^m u$ are as in the real case themselves infinitely differentiable. The operators Q_ℓ^m can be extended from the space of the infinitely differentiable functions with compact support to L_2 and represent then L_2-orthogonal projections onto subspaces of L_2 of the same structure as in the real case.

Lemma. *If $u : \mathbb{R}^3 \to \mathbb{C}$ is an infinitely differentiable function,*

$$u = Q_\ell^m u \quad \Leftrightarrow \quad L^2 u = \ell(\ell+1)u, \quad L_3 u = mu.$$

Proof. For all such functions u and all admissible indices ℓ and m,

$$L^2 Q_\ell^m u = \ell(\ell+1) Q_\ell^m u, \quad L_3 Q_\ell^m u = m Q_\ell^m u,$$

which can be shown as in the proof of Lemma 9.2 and uses that H_ℓ^m is a joint eigenfunction of the operators L^2 and L_3. If therefore $u = Q_\ell^m u$, then

$$L^2 u = \ell(\ell+1)u, \quad L_3 u = mu.$$

Conversely let u be an infinitely differentiable function that satisfies these equations. For all infinitely differentiable functions v with compact support,

$$\ell'(\ell'+1)(u, Q_{\ell'}^{m'} v) = (u, L^2 Q_{\ell'}^{m'} v) = (L^2 u, Q_{\ell'}^{m'} v) = \ell(\ell+1)(u, Q_{\ell'}^{m'} v),$$

$$m'(u, Q_{\ell'}^{m'} v) = (u, L_3 Q_{\ell'}^{m'} v) = (L_3 u, Q_{\ell'}^{m'} v) = m(u, Q_{\ell'}^{m'} v).$$

That is, $(u, Q_{\ell'}^{m'} v) = 0$ if $\ell' \neq \ell$ or $m' \neq m$. By the complex version of Theorem 9.1

$$(u, v) = \sum_{\ell'=0}^{\infty} \sum_{m'=-\ell'}^{\ell'} (u, Q_{\ell'}^{m'} v) = (u, Q_{\ell}^{m} v)$$

follows. This implies $(u, v) = (Q_{\ell}^{m} u, v)$ and, as v was arbitrary, $u = Q_{\ell}^{m} u$. $\qquad\square$

The Fourier transform commutes with the operators L^2 and L_3:

Lemma. *For all infinitely differentiable functions u with compact support,*

$$(L_\nu \widehat{u})(\omega) = \left(\frac{1}{\sqrt{2\pi}}\right)^3 \int (L_\nu u)(x) e^{-i\omega \cdot x} dx, \quad \nu = 1, 2, 3,$$

or, in abbreviated form, $L_\nu F u = F L_\nu u$. Moreover $L^2 F u = F L^2 u$.

Proof. The proof is based on integration by parts. For example,

$$-i\left(\omega_2 \frac{\partial}{\partial \omega_3} - \omega_3 \frac{\partial}{\partial \omega_2}\right) \widehat{u}(\omega)$$

$$= -i\left(\frac{1}{\sqrt{2\pi}}\right)^3 \int u(x)\left(-ix_3\,\omega_2 + i\omega_3 x_2\right) e^{-i\omega \cdot x} dx$$

$$= -i\left(\frac{1}{\sqrt{2\pi}}\right)^3 \int u(x)\left(x_3 \frac{\partial}{\partial x_2} - x_2 \frac{\partial}{\partial x_3}\right)\left\{e^{-i\omega \cdot x}\right\} dx$$

$$= -i\left(\frac{1}{\sqrt{2\pi}}\right)^3 \int \left(x_2 \frac{\partial}{\partial x_3} - x_3 \frac{\partial}{\partial x_2}\right)\left\{u(x)\right\} e^{-i\omega \cdot x} dx$$

and therefore $L_1 F u = F L_1 u$. The other components are treated analogously. $\qquad\square$

The Fourier transform commutes therefore, for the given basis of the spherical harmonics, also with the radial-angular decomposition from Theorem 9.1:

Theorem. *For all square integrable functions $u : \mathbb{R}^3 \to \mathbb{C}$,*

$$Q_{\ell}^{m} F u = F Q_{\ell}^{m} u.$$

Proof. Let u be first an infinitely differentiable function with compact support. Its projections $Q_{\ell'}^{m'} u$ are then of the same type and the following identities hold:

$$L^2 F Q_{\ell'}^{m'} u = F L^2 Q_{\ell'}^{m'} u = \ell'(\ell'+1) F Q_{\ell'}^{m'} u,$$

$$L_3 F Q_{\ell'}^{m'} u = F L_3 Q_{\ell'}^{m'} u = m' F Q_{\ell'}^{m'} u.$$

By the Lemma above therefore $F Q_{\ell'}^{m'} u = Q_{\ell'}^{m'} F Q_{\ell'}^{m'} u$. This implies $Q_{\ell}^{m} F Q_{\ell'}^{m'} u = 0$ if $\ell' \neq \ell$ or $m' \neq m$ and thus, by the continuity of the operator $Q_{\ell}^{m} F$ and Theorem 9.1,

$$Q_\ell^m F u = \sum_{\ell'=0}^{\infty} \sum_{m'=-\ell'}^{\ell'} Q_\ell^m F Q_{\ell'}^{m'} u = F Q_\ell^m u.$$

Since the infinitely differentiable functions with compact support are dense in L_2 and the operators F and Q_ℓ^m from L_2 to L_2 are bounded, the proposition follows. $\square$

This means particularly that the Fourier transform Fu of a square integrable function coincides with its projection $Q_\ell^m F u$ if and only if $u = Q_\ell^m u$. The Fourier transform and the Fourier back transform of a square integrable function

$$x \to \frac{1}{r} f(r) K_\ell^m(x), \quad r = |x|,$$

are therefore of the same form as the function itself.

References

1. Abramowitz, M., Stegun, I.: Handbook of Mathematical Functions. Dover Publications, New York (10th printing in 1972)
2. Adams, R., Fournier, J.: Sobolev Spaces. Elsevier, Amsterdam (2003)
3. Agmon, S.: Lectures on the Exponential Decay of Solutions of Second-Order Elliptic Operators. Princeton University Press, Princeton (1981)
4. Ahlrichs, R.: Asymptotic behavior of atomic bound state wavefunctions. J. Math. Phys. **14**, 1860–1863 (1973)
5. Aigner, M.: A Course in Enumeration. Springer, Berlin Heidelberg New York (2007)
6. Atkins, P., Friedman, R.: Molecular Quantum Mechanics. Oxford University Press, Oxford (1997)
7. Babenko, K.: Approximation by trigonometric polynomials in a certain class of periodic functions of several variables. Sov. Math., Dokl. **1**, 672–675 (1960)
8. Babuška, I., Osborn, J.: Finite element-Galerkin approximation of the eigenvalues and eigenvectors of selfadjoint problems. Math. Comput. **52**, 275–297 (1989)
9. Babuška, I., Osborn, J.: Eigenvalue problems. In: P. Ciarlet, J. Lions (eds.) Handbook of Numerical Analysis, Vol. II, Finite Element Methods (Part 1), pp. 641–792. Elsevier, Amsterdam (1991)
10. Beylkin, G., Mohlenkamp, M., Perez, F.: Approximating a wavefunction as an unconstrained sum of Slater determinants. J. Math. Phys. **49**, 032,107 (2008)
11. Born, M., Oppenheimer, R.: Zur Quantentheorie der Molekeln. Ann. der Physik **84**, 457–484 (1927)
12. Boys, S.: A general method of calculation for the stationary states of any molecular system. Proc. R. Soc. Lond., Ser. A, Math. Phys. Eng. Sci. **200**, 542–554 (1950)
13. Braess, D.: Asymptotics for the approximation of wave functions by exponential sums. J. Approx. Theory **83**, 93–103 (1995)
14. Bramble, J., Pasciak, J., Steinbach, O.: On the stability of the L_2-projection in $H^1(\Omega)$. Math. Comput. **71**, 147–156 (2001)
15. Bungartz, H.J., Griebel, M.: Sparse grids. Acta Numerica **13**, 1–123 (2004)
16. Cancès, E., Le Bris, C., Maday, Y.: Méthodes Mathématiques en Chimie Quantique. Springer, Berlin Heidelberg New York (2006)
17. Carstensen, C.: Merging the Bramble-Pasciak-Steinbach and the Crouzeix-Thomée criterion for H^1-stability of the L_2-projection onto finite element spaces. Math. Comput. **71**, 157–163 (2001)
18. Cohen-Tannoudji, C., Diu, B., Laloë, F.: Quantum Mechanics, vols. I, II. John Wiley & Sons, New York (1977)
19. Combes, J., Thomas, L.: Asymptotic behavior of eigenfunctions for multiparticle Schrödinger operators. Commun. Math. Phys. **34**, 251–270 (1973)
20. Connor, A.O.: Exponential decay of bound state wave functions. Commun. Math. Phys. **32**, 319–340 (1973)
21. Courant, R., Hilbert, D.: Methoden der Mathematischen Physik I. Springer, Berlin (1924)

22. Deift, P., Hunziker, W., Simon, B., Vock, E.: Pointwise bounds on eigenfunctions and wave packets in N-body quantum systems IV. Commun. Math. Phys. **64**, 1–34 (1978)
23. Delvos, F.: d-variate Boolean interpolation. J. Approx. Theory **34**, 99–114 (1982)
24. Delvos, F., Schempp, W.: Boolean methods in Interpolation and Approximation. Pitman Research Notes in Mathematics, vol. 230. John Wiley & Sons, New York (1989)
25. Dirac, P.: Quantum mechanics of many electron systems. Proc. R. Soc. Lond., Ser. A, Math. Phys. Eng. Sci. **123**, 714–733 (1929)
26. Euler, L.: De partitione numerorum. Novi commentarii academiae scientarium Petropolitanae **3**, 125–169 (1753). Reprinted in Opera omnia, Series I, vol. 2, pp. 254–294
27. Flad, H.J., Hackbusch, W., Kolb, D., Schneider, R.: Wavelet approximation of correlated wave functions. I. Basics. J. Chem. Phys. **116**, 9461–9657 (2002)
28. Flad, H.J., Hackbusch, W., Kolb, D., Koprucki, T.: Wavelet approximation of correlated wave functions. II. Hyperbolic wavelets and adaptive approximation schemes. J. Chem. Phys. **117**, 3625–3638 (2002)
29. Flad, H.J., Hackbusch, W., Schneider, R.: Best N-term approximation in electronic structure calculations. I. One-electron reduced density matrix. M2AN **40**, 49–61 (2006)
30. Flad, H.J., Hackbusch, W., Schneider, R.: Best N-term approximation in electronic structure calculations. II. Jastrow factors. M2AN **41**, 261–279 (2007)
31. Flad, H.J., Schneider, R., Schulze, B.W.: Asymptotic regularity of solutions of Hartree-Fock equations with Coulomb potential. Math. Methods Appl. Sci. **31**, 2172–2201 (2008)
32. Fournais, S., Hoffmann-Ostenhof, M., Hoffmann-Ostenhof, T., Østergard Sørensen, T.: Sharp regularity estimates for Coulombic many-electron wave functions. Commun. Math. Phys. **255**, 183–227 (2005)
33. Friesecke, G.: The multiconfiguration equations for atoms and molecules: charge quantization and existence of solutions. Arch. Ration. Mech. Anal. **169**, 35–71 (2003)
34. Garcke, J., Griebel, M.: On the computation of the eigenproblems of hydrogen and helium in strong magnetic and electric fields with the sparse grid combination technique. J. Comput. Phys. **165**, 694–716 (2000)
35. Griebel, M.: Sparse grids and related approximation schemes for higher dimensional problems. In: Foundations of Computational Mathematics, Santander 2005, Lond. Math. Soc. Lect. Note Ser., vol. 331, pp. 106–161. Cambridge University Press, Cambridge (2006)
36. Griebel, M., Hamaekers, J.: A wavelet based sparse grid method for the electronic Schrödinger equation. In: International Congress of Mathematicians, vol. III, pp. 1473–1506. Eur. Math. Soc., Zürich (2006)
37. Griebel, M., Hamaekers, J.: Sparse grids for the Schrödinger equation. M2AN **41**, 215–247 (2007)
38. Gustafson, S., Sigal, I.: Mathematical Concepts of Quantum Mechanics. Springer, Berlin Heidelberg New York (2003)
39. Hackbusch, W.: The efficient computation of certain determinants arising in the treatment of Schrödinger's equation. Computing **67**, 35–56 (2000)
40. Hamaekers, J.: Tensor Product Multiscale Many-Particle Spaces with Finite-Order Weights for the Electronic Schrödinger Equation. Doctoral thesis, Universität Bonn (2009)
41. Hardy, G., Ramanujan, S.: Asymptotic formulae in combinatory analysis. Proc. Lond. Math. Soc. **17**, 75–115 (1918)
42. Helgaker, T., Jørgensen, P., Olsen, J.: Molecular Electronic Structure Theory. John Wiley & Sons, Chichester (2000)
43. Hilgenfeldt, S., Balder, S., Zenger, C.: Sparse grids: applications to multi-dimensional Schrödinger problems. SFB-Bericht 342/05/95, TU München, München (1995)
44. Hislop, P., Sigal, I.: Introduction to Spectral Theory with Applications to Schrödinger Operators. Springer, Berlin Heidelberg New York (1996)
45. Hoffmann-Ostenhof, M., Hoffmann-Ostenhof, T., Stremnitzer, H.: Local properties of Coulombic wave functions. Commun. Math. Phys. **163**, 185–215 (1994)
46. Hunziker, W.: On the spectra of Schrödinger multiparticle Hamiltonians. Helv. Phys. Acta **39**, 451–462 (1966)

47. Hunziker, W., Sigal, I.: The quantum N-body problem. J. Math. Phys. **41**, 3448–3510 (2000)
48. Kato, T.: Fundamental properties of Hamiltonian operators of Schrödinger type. Trans. Am. Math. Soc. **70**, 195–221 (1951)
49. Kato, T.: On the eigenfunctions of many-particle systems in quantum mechanics. Commun. Pure Appl. Math. **10**, 151–177 (1957)
50. Knyazev, A., Osborn, J.: New a priori FEM error estimates for eigenvalues. SIAM J. Numer. Anal. **43**, 2647–2667 (2006)
51. Kohn, W.: Nobel lecture: Electronic structure of matter-wave functions and density functionals. Rev. Mod. Phys. **71**, 1253–1266 (1999)
52. Königsberger, K.: Analysis 1. Springer, Berlin Heidelberg New York (2004)
53. Korobov, N.: Approximate calculation of repeated integrals by number-theoretical methods (Russian). Dokl. Akad. Nauk. SSSR **115**, 1062–1065 (1957)
54. Korobov, N.: Approximate calculation of repeated integrals (Russian). Dokl. Akad. Nauk. SSSR **124**, 1207–1210 (1959)
55. Kutzelnigg, W.: Theory of the expansion of wave functions in a Gaussian basis. International Journal of Quantum Chemistry **51**, 447–463 (1994)
56. Le Bris, C. (ed.): Handbook of Numerical Analysis, Vol. X: Computational Chemistry. North Holland, Amsterdam (2003)
57. Le Bris, C.: Computational chemistry from the perspective of numerical analysis. Acta Numerica **14**, 363–444 (2005)
58. Le Bris, C., Lions, P.: From atoms to crystals: a mathematical journey. Bull. Am. Math. Soc., New Ser. **42**, 291–363 (2005)
59. Lewin, M.: Solutions of the multiconfiguration equations in quantum chemistry. Arch. Ration. Mech. Anal. **171**, 83–114 (2004)
60. Lieb, E., Simon, B.: The Hartree-Fock theory for Coulomb systems. Commun. Math. Phys. **53**, 185–194 (1977)
61. Lieb, E., Simon, B.: The Thomas-Fermi theory of atoms, molecules, and solids. Adv. Math. **23**, 22–116 (1977)
62. Lions, P.: Solutions of Hartree-Fock equations for Coulomb systems. Commun. Math. Phys. **109**, 33–97 (1987)
63. Messiah, A.: Quantum Mechanics. Dover Publications, New York (2000)
64. von Neumann, J.: Mathematische Grundlagen der Quantenmechanik. Springer, Berlin (1932)
65. Persson, A.: Bounds for the discrete part of the spectrum of a semi-bounded Schrödinger operator. Math. Scand. **8**, 143–153 (1960)
66. Pople, J.: Nobel lecture: Quantum chemical models. Rev. Mod. Phys. **71**, 1267–1274 (1999)
67. Rademacher, H.: On the partition function $p(n)$. Proc. Lond. Math. Soc. **43**, 241–254 (1937)
68. Raviart, P., Thomas, J.: Introduction à L'Analyse Numérique des Équations aux Dérivées Partielles. Masson, Paris (1983)
69. Reed, M., Simon, B.: Methods of Modern Mathematical Physics I: Functional Analysis. Academic Press, San Diego (1980)
70. Reed, M., Simon, B.: Methods of Modern Mathematical Physics II: Fourier Analysis, Self Adjointness. Academic Press, San Diego (1975)
71. Reed, M., Simon, B.: Methods of Modern Mathematical Physics IV: Analysis of Operators. Academic Press, San Diego (1978)
72. Schneider, R.: Analysis of the projected coupled cluster method in electronic structure calculation. Numer. Math. **113**, 433–471 (2009)
73. Schrödinger, E.: Quantisierung als Eigenwertproblem. Ann. der Physik **79**, 361–376 (1926)
74. Simon, B.: Pointwise bounds on eigenfunctions and wave packets in N-body quantum system I. Proc. Am. Math. Soc. **208**, 317–329 (1975)
75. Simon, B.: Schrödinger operators in the twentieth century. J. Math. Phys. **41**, 3523–3555 (2000)
76. Smolyak, S.: Quadrature and interpolation formulas for tensor products of certain classes of functions. Dokl. Akad. Nauk SSSR **4**, 240–243 (1963)
77. Stein, E., Weiss, G.: Introduction to Fourier Analysis on Euclidean Spaces. Princeton University Press, Princeton (1971)

78. Teufel, S.: Adiabatic Perturbation Theory in Quantum Dynamics. Lecture Notes in Mathematics 1821. Springer, Berlin Heidelberg New York (2003)
79. Thaller, B.: Visual Quantum Mechanics. Springer, New York (2000)
80. Thaller, B.: Advanced Visual Quantum Mechanics. Springer, New York (2004)
81. Titchmarsh, E.: Eigenfunction Expansions, vols. I and II. Oxford University Press, Oxford (1953, 1958)
82. Titchmarsh, E.: On the eigenvalues in problems with spherical symmetry. Proc. R. Soc. Lond., Ser. A, Math. Phys. Eng. Sci. **245**, 147–155 (1958)
83. Titchmarsh, E.: On the eigenvalues in problems with spherical symmetry II. Proc. R. Soc. Lond., Ser. A, Math. Phys. Eng. Sci. **251**, 46–54 (1959)
84. Titchmarsh, E.: On the eigenvalues in problems with spherical symmetry III. Proc. R. Soc. Lond., Ser. A, Math. Phys. Eng. Sci. **252**, 436–444 (1959)
85. Triebel, H.: Theory of Function Spaces. Birkhäuser, Basel (1983)
86. Weidmann, J.: Linear Operators in Hilbert Spaces. Springer, New York Heidelberg Berlin (1980)
87. Weidmann, J.: Lineare Operatoren in Hilberträumen, Teil I, Grundlagen. B.G. Teubner, Stuttgart Leipzig Wiesbaden (2000)
88. Weidmann, J.: Lineare Operatoren in Hilberträumen, Teil II, Anwendungen. B.G. Teubner, Stuttgart Leipzig Wiesbaden (2003)
89. Weyl, H.: Das asymptotische Verteilungsgesetz der Eigenwerte linearer partieller Differentialgleichungen. Math. Ann. **71**, 441–479 (1912)
90. van Winter, C.: Theory of finite systems of particles. Mat.-Fys. Skr. Danske Vid. Selsk. **1, 2** (1964, 1965)
91. Yoshida, K.: Functional Analysis. Die Grundlehren der mathematischen Wissenschaften, vol.123. Springer, Berlin Heidelberg New York (1971)
92. Yserentant, H.: On the regularity of the electronic Schrödinger equation in Hilbert spaces of mixed derivatives. Numer. Math. **98**, 731–759 (2004)
93. Yserentant, H.: Sparse grid spaces for the numerical solution of the electronic Schrödinger equation. Numer. Math. **101**, 381–389 (2005)
94. Yserentant, H.: The hyperbolic cross space approximation of electronic wavefunctions. Numer. Math. **105**, 659–690 (2007)
95. Yserentant, H.: Regularity properties of wavefunctions and the complexity of the quantum mechanical N-body problem. Unpublished manuscript (November 2007)
96. Zeiser, A.: Direkte Diskretisierung der Schrödingergleichung auf dünnen Gittern. Doctoral thesis, Technische Universität Berlin, in preparation
97. Zenger, C.: Sparse grids. In: W. Hackbusch (ed.) Parallel Algorithms for Partial Differential Equations, Kiel 1990, Notes on Numerical Fluid Mechanics, vol. 31, pp. 241–251. Vieweg, Braunschweig Wiesbaden (1991)
98. Zhislin, G.: A study of the spectrum of the Schrödinger operator for a system of several particles (Russian). Tr. Mosk. Mat. O.-va **9**, 81–120 (1960)
99. Ziemer, W.: Weakly Differentiable Functions. Springer, New York Heidelberg Berlin (1989)

Index

Lecture Notes in Mathematics

For information about earlier volumes
please contact your bookseller or Springer
LNM Online archive: springerlink.com

Vol. 1857: M. Émery, M. Ledoux, M. Yor (Eds.), Séminaire de Probabilités XXXVIII (2005)

Vol. 1858: A.S. Cherny, H.-J. Engelbert, Singular Stochastic Differential Equations (2005)

Vol. 1859: E. Letellier, Fourier Transforms of Invariant Functions on Finite Reductive Lie Algebras (2005)

Vol. 1860: A. Borisyuk, G.B. Ermentrout, A. Friedman, D. Terman, Tutorials in Mathematical Biosciences I. Mathematical Neurosciences (2005)

Vol. 1861: G. Benettin, J. Henrard, S. Kuksin, Hamiltonian Dynamics – Theory and Applications, Cetraro, Italy, 1999. Editor: A. Giorgilli (2005)

Vol. 1862: B. Helffer, F. Nier, Hypoelliptic Estimates and Spectral Theory for Fokker-Planck Operators and Witten Laplacians (2005)

Vol. 1863: H. Führ, Abstract Harmonic Analysis of Continuous Wavelet Transforms (2005)

Vol. 1864: K. Efstathiou, Metamorphoses of Hamiltonian Systems with Symmetries (2005)

Vol. 1865: D. Applebaum, B.V. R. Bhat, J. Kustermans, J. M. Lindsay, Quantum Independent Increment Processes I. From Classical Probability to Quantum Stochastic Calculus. Editors: M. Schürmann, U. Franz (2005)

Vol. 1866: O.E. Barndorff-Nielsen, U. Franz, R. Gohm, B. Kümmerer, S. Thorbjønsen, Quantum Independent Increment Processes II. Structure of Quantum Lévy Processes, Classical Probability, and Physics. Editors: M. Schürmann, U. Franz, (2005)

Vol. 1867: J. Sneyd (Ed.), Tutorials in Mathematical Biosciences II. Mathematical Modeling of Calcium Dynamics and Signal Transduction. (2005)

Vol. 1868: J. Jorgenson, S. Lang, $Pos_n(R)$ and Eisenstein Series. (2005)

Vol. 1869: A. Dembo, T. Funaki, Lectures on Probability Theory and Statistics. Ecole d'Eté de Probabilités de Saint-Flour XXXIII-2003. Editor: J. Picard (2005)

Vol. 1870: V.I. Gurariy, W. Lusky, Geometry of Müntz Spaces and Related Questions. (2005)

Vol. 1871: P. Constantin, G. Gallavotti, A.V. Kazhikhov, Y. Meyer, S. Ukai, Mathematical Foundation of Turbulent Viscous Flows, Martina Franca, Italy, 2003. Editors: M. Cannone, T. Miyakawa (2006)

Vol. 1872: A. Friedman (Ed.), Tutorials in Mathematical Biosciences III. Cell Cycle, Proliferation, and Cancer (2006)

Vol. 1873: R. Mansuy, M. Yor, Random Times and Enlargements of Filtrations in a Brownian Setting (2006)

Vol. 1874: M. Yor, M. Émery (Eds.), In Memoriam Paul-André Meyer - Séminaire de Probabilités XXXIX (2006)

Vol. 1875: J. Pitman, Combinatorial Stochastic Processes. Ecole d'Eté de Probabilités de Saint-Flour XXXII-2002. Editor: J. Picard (2006)

Vol. 1876: H. Herrlich, Axiom of Choice (2006)

Vol. 1877: J. Steuding, Value Distributions of L-Functions (2007)

Vol. 1878: R. Cerf, The Wulff Crystal in Ising and Percolation Models, Ecole d'Eté de Probabilités de Saint-Flour XXXIV-2004. Editor: Jean Picard (2006)

Vol. 1879: G. Slade, The Lace Expansion and its Applications, Ecole d'Eté de Probabilités de Saint-Flour XXXIV-2004. Editor: Jean Picard (2006)

Vol. 1880: S. Attal, A. Joye, C.-A. Pillet, Open Quantum Systems I, The Hamiltonian Approach (2006)

Vol. 1881: S. Attal, A. Joye, C.-A. Pillet, Open Quantum Systems II, The Markovian Approach (2006)

Vol. 1882: S. Attal, A. Joye, C.-A. Pillet, Open Quantum Systems III, Recent Developments (2006)

Vol. 1883: W. Van Assche, F. Marcellàn (Eds.), Orthogonal Polynomials and Special Functions, Computation and Application (2006)

Vol. 1884: N. Hayashi, E.I. Kaikina, P.I. Naumkin, I.A. Shishmarev, Asymptotics for Dissipative Nonlinear Equations (2006)

Vol. 1885: A. Telcs, The Art of Random Walks (2006)

Vol. 1886: S. Takamura, Splitting Deformations of Degenerations of Complex Curves (2006)

Vol. 1887: K. Habermann, L. Habermann, Introduction to Symplectic Dirac Operators (2006)

Vol. 1888: J. van der Hoeven, Transseries and Real Differential Algebra (2006)

Vol. 1889: G. Osipenko, Dynamical Systems, Graphs, and Algorithms (2006)

Vol. 1890: M. Bunge, J. Funk, Singular Coverings of Toposes (2006)

Vol. 1891: J.B. Friedlander, D.R. Heath-Brown, H. Iwaniec, J. Kaczorowski, Analytic Number Theory, Cetraro, Italy, 2002. Editors: A. Perelli, C. Viola (2006)

Vol. 1892: A. Baddeley, I. Bárány, R. Schneider, W. Weil, Stochastic Geometry, Martina Franca, Italy, 2004. Editor: W. Weil (2007)

Vol. 1893: H. Hanßmann, Local and Semi-Local Bifurcations in Hamiltonian Dynamical Systems, Results and Examples (2007)

Vol. 1894: C.W. Groetsch, Stable Approximate Evaluation of Unbounded Operators (2007)

Vol. 1895: L. Molnár, Selected Preserver Problems on Algebraic Structures of Linear Operators and on Function Spaces (2007)

Vol. 1896: P. Massart, Concentration Inequalities and Model Selection, Ecole d'Été de Probabilités de Saint-Flour XXXIII-2003. Editor: J. Picard (2007)

Vol. 1897: R. Doney, Fluctuation Theory for Lévy Processes, Ecole d'Été de Probabilités de Saint-Flour XXXV-2005. Editor: J. Picard (2007)

Vol. 1898: H.R. Beyer, Beyond Partial Differential Equations, On linear and Quasi-Linear Abstract Hyperbolic Evolution Equations (2007)

Vol. 1899: Séminaire de Probabilités XL. Editors: C. Donati-Martin, M. Émery, A. Rouault, C. Stricker (2007)

Vol. 1900: E. Bolthausen, A. Bovier (Eds.), Spin Glasses (2007)

Vol. 1901: O. Wittenberg, Intersections de deux quadriques et pinceaux de courbes de genre 1, Intersections of Two Quadrics and Pencils of Curves of Genus 1 (2007)

Vol. 1902: A. Isaev, Lectures on the Automorphism Groups of Kobayashi-Hyperbolic Manifolds (2007)

Vol. 1903: G. Kresin, V. Maz'ya, Sharp Real-Part Theorems (2007)

Vol. 1904: P. Giesl, Construction of Global Lyapunov Functions Using Radial Basis Functions (2007)

Vol. 1905: C. Prévôt, M. Röckner, A Concise Course on Stochastic Partial Differential Equations (2007)

Vol. 1906: T. Schuster, The Method of Approximate Inverse: Theory and Applications (2007)

Vol. 1907: M. Rasmussen, Attractivity and Bifurcation for Nonautonomous Dynamical Systems (2007)

Vol. 1908: T.J. Lyons, M. Caruana, T. Lévy, Differential Equations Driven by Rough Paths, Ecole d'Été de Probabilités de Saint-Flour XXXIV-2004 (2007)

Recent Reprints and New Editions

LECTURE NOTES IN MATHEMATICS  **Springer**

Edited by J.-M. Morel, F. Takens, B. Teissier, P.K. Maini

Editorial Policy (for the publication of monographs)

1. Lecture Notes aim to report new developments in all areas of mathematics and their applications - quickly, informally and at a high level. Mathematical texts analysing new developments in modelling and numerical simulation are welcome.

   Monograph manuscripts should be reasonably self-contained and rounded off. Thus they may, and often will, present not only results of the author but also related work by other people. They may be based on specialised lecture courses. Furthermore, the manuscripts should provide sufficient motivation, examples and applications. This clearly distinguishes Lecture Notes from journal articles or technical reports which normally are very concise. Articles intended for a journal but too long to be accepted by most journals, usually do not have this "lecture notes" character. For similar reasons it is unusual for doctoral theses to be accepted for the Lecture Notes series, though habilitation theses may be appropriate.

2. Manuscripts should be submitted either online at www.editorialmanager.com/lnm to Springer's mathematics editorial in Heidelberg, or to one of the series editors. In general, manuscripts will be sent out to 2 external referees for evaluation. If a decision cannot yet be reached on the basis of the first 2 reports, further referees may be contacted: The author will be informed of this. A final decision to publish can be made only on the basis of the complete manuscript, however a refereeing process leading to a preliminary decision can be based on a pre-final or incomplete manuscript. The strict minimum amount of material that will be considered should include a detailed outline describing the planned contents of each chapter, a bibliography and several sample chapters.

   Authors should be aware that incomplete or insufficiently close to final manuscripts almost always result in longer refereeing times and nevertheless unclear referees' recommendations, making further refereeing of a final draft necessary.

   Authors should also be aware that parallel submission of their manuscript to another publisher while under consideration for LNM will in general lead to immediate rejection.

3. Manuscripts should in general be submitted in English. Final manuscripts should contain at least 100 pages of mathematical text and should always include

   - a table of contents;
   - an informative introduction, with adequate motivation and perhaps some historical remarks: it should be accessible to a reader not intimately familiar with the topic treated;
   - a subject index: as a rule this is genuinely helpful for the reader.

   For evaluation purposes, manuscripts may be submitted in print or electronic form (print form is still preferred by most referees), in the latter case preferably as pdf- or zipped ps-files. Lecture Notes volumes are, as a rule, printed digitally from the authors' files. To ensure best results, authors are asked to use the LaTeX2e style files available from Springer's web-server at:

   ftp://ftp.springer.de/pub/tex/latex/svmonot1/ (for monographs) and
   ftp://ftp.springer.de/pub/tex/latex/svmultt1/ (for summer schools/tutorials).

Additional technical instructions, if necessary, are available on request from: lnm@springer.com.

4. Careful preparation of the manuscripts will help keep production time short besides ensuring satisfactory appearance of the finished book in print and online. After acceptance of the manuscript authors will be asked to prepare the final LaTeX source files and also the corresponding dvi-, pdf- or zipped ps-file. The LaTeX source files are essential for producing the full-text online version of the book (see http://www.springerlink.com/openurl.asp?genre=journal&issn=0075-8434 for the existing online volumes of LNM).

   The actual production of a Lecture Notes volume takes approximately 12 weeks.

5. Authors receive a total of 50 free copies of their volume, but no royalties. They are entitled to a discount of 33.3% on the price of Springer books purchased for their personal use, if ordering directly from Springer.

6. Commitment to publish is made by letter of intent rather than by signing a formal contract. Springer-Verlag secures the copyright for each volume. Authors are free to reuse material contained in their LNM volumes in later publications: a brief written (or e-mail) request for formal permission is sufficient.

Addresses:
Professor J.-M. Morel, CMLA,
École Normale Supérieure de Cachan,
61 Avenue du Président Wilson, 94235 Cachan Cedex, France
E-mail: Jean-Michel.Morel@cmla.ens-cachan.fr

Professor F. Takens, Mathematisch Instituut,
Rijksuniversiteit Groningen, Postbus 800,
9700 AV Groningen, The Netherlands
E-mail: F.Takens@rug.nl

Professor B. Teissier, Institut Mathématique de Jussieu,
UMR 7586 du CNRS, Équipe "Géométrie et Dynamique",
175 rue du Chevaleret,
75013 Paris, France
E-mail: teissier@math.jussieu.fr

For the "Mathematical Biosciences Subseries" of LNM:

Professor P.K. Maini, Center for Mathematical Biology,
Mathematical Institute, 24-29 St Giles,
Oxford OX1 3LP, UK
E-mail: maini@maths.ox.ac.uk

Springer, Mathematics Editorial, Tiergartenstr. 17,
69121 Heidelberg, Germany,
Tel.: +49 (6221) 487-259
Fax: +49 (6221) 4876-8259
E-mail: lnm@springer.com